NEW SPACE FRONTIERS

First published in 2014 by Zenith Press, an imprint of Quarto Publishing Group USA Inc., 400 First Avenue North, Suite 400, Minneapolis, MN 55401 USA

Zenith Press titles are also available at discounts in bulk quantity for industrial or sales-promotional use. For details write to Special Sales Manager at Quarto Publishing Group USA Inc., 400 First Avenue North, Suite 400, Minneapolis, MN 55401 USA.

To find out more about our books, visit us online at www.zenithpress.com.

ISBN: 978-0-7603-4666-2

Library of Congress Cataloging-in-Publication Data

Bizony, Piers, author.
 New space frontiers : venturing into Earth orbit and beyond / by Piers Bizony.
   pages cm
  ISBN 978-0-7603-4666-2 (hardback)
 1. Space flight--Technological innovations--Forecasting. 2. Outer space--Exploration--Forecasting. 3. Astronautics and state--Forecastng. I. Title.
 TL790.B56 2014
 629.4'10112--dc23
                          2014018427

Layout: Spaced Design Ltd.

On the front cover: *NASA/Sierra Nevada Corporation*
On the back cover: *World View Enterprises, Inc.*

Printed in China

10 9 8 7 6 5 4 3 2 1

NEW SPACE FRONTIERS

FRONTIERS

venturing into earth orbit and beyond

ZENITH PRESS

milestones in human space flight

Russian pilot **Yuri Gagarin** becomes the first man in space, completing one orbit of the Earth aboard a Vostok capsule.
APRIL 12, 1961

John Glenn makes the first U.S. manned orbital flight aboard Mercury 6, completing three orbits before reentry and splashdown.
FEBRUARY 20, 1962

Soviet cosmonaut **Alexei Leonov** makes the first spacewalk from Voskhod 2, a Vostok-style craft adapted for two crewmen.
MARCH 18, 1965

Gemini 8 links with a previously launched unmanned Agena target vehicle, making the first firm orbital docking between two spacecraft.
MARCH 16, 1966

The Soyuz 1 spacecraft is launched, but multiple failures cause the death of Soviet cosmonaut **Vladimir Komarov**.
APRIL 23, 1967

Two Soviet spacecraft, Soyuz 4 and Soyuz 5, rendezvous and dock. Crews **transfer between ships** in orbit for the first time.
JANUARY 16, 1969

An explosion damages **Apollo 13**. The astronauts use the lunar module to slingshot around the Moon to speed their return to Earth.
APRIL 13, 1970

Soyuz 11 launches successfully, docking with Salyut 1, but the three cosmonauts are **killed during reentry** because of an air leak in the cabin.
JUNE 6, 1971

MAY 5, 1961
NASA's "Freedom 7" Mercury capsule launches on a Redstone rocket for a suborbital flight, making **Alan Shepard** the first American in space.

JUNE 16, 1963
Valentina Tereshkova of the U.S.S.R. becomes the first woman to fly into space, aboard a Vostok capsule.

JUNE 3, 1965
Ed White becomes the first American to walk in space during Gemini 4. He floats almost free of the capsule, attached by an umbilical cable.

JANUARY 27, 1967
Gus Grissom, Roger Chaffee, and Ed White of Apollo 1 suffocate from smoke inhalation when their capsule catches fire during a launch pad test.

DECEMBER 21, 1968
Apollo 8 launches on a Saturn V prior to becoming the first manned mission to fly into deep space and orbit the Moon.

JULY 20, 1969
Apollo 11's lunar module Eagle crew lands on the Moon. Neil Armstrong and Buzz Aldrin walk on the lunar surface.

APRIL 19, 1971
A Proton rocket launches the **first small space station**, the U.S.S.R.'s Salyut 1, from the Soviet-controlled "cosmodrome" at Baikonur in Kazakhstan.

APRIL 12, 1981
Space Shuttle Columbia lifts off from Cape Canaveral, beginning the **first orbital mission** for NASA's new rocket plane.

JANUARY 28, 1986
Shuttle *Challenger* explodes 73 seconds after launch, killing the crew of seven and grounding the Shuttle fleet for more than two years.

DECEMBER 2, 1993
Astronauts aboard *Endeavour* lift off for a mission to repair and upgrade the **Hubble Space Telescope**, bringing it to full working condition.

MAY 6, 2001
American entrepreneur **Dennis Tito** returns to Earth aboard a Russian Soyuz spacecraft after becoming the first privately funded space voyager.

OCTOBER 15, 2003
Yang Liwei becomes China's first "taikonaut" aboard the *Shenzhou* spacecraft. China becomes the third nation capable of human space flight.

JULY 8, 2011
Shuttle *Atlantis* carrying supplies for the International Space Station, launches for the 135th and **final mission** of the Space Shuttle era.

MAY 29, 2014
Elon Musk's **SpaceX** company unveils the first privately built crew transport capsule, the Dragon V2, due for first flight in 2017.

MAY 14, 1973
A Saturn V rocket launches **Skylab**, NASA's first space station, built from an adapted Saturn V stage. Three crews visit between 1973 and 1974.

JUNE 18, 1983
NASA astronaut **Sally Ride** becomes the first American woman to fly into orbit aboard Space Shuttle *Challenger.*

FEBRUARY 20, 1986
The Soviet Union launches the first element of the multi-module **Mir space station**, which would remain operational for fourteen years.

NOVEMBER 20, 1998
Russia's *Zarya* control module, the first segment of the **International Space Station**, is launched into orbit.

FEBRUARY 1, 2003
Shuttle *Columbia* disintegrates as it re-enters the atmosphere. All crew members are lost. Loose insulation had punctured a wing during lift-off.

OCTOBER 4, 2004
Scaled Composites's *SpaceShipOne* piloted rocket plane wins the X Prize by flying into suborbital space twice within two weeks.

JUNE 29, 2012
The first **Orion spacecraft** arrives at NASA's Kennedy Space Center for final checks, heralding a new era in human space flight.

contents

01.

1 ESCAPE FROM **PLANET EARTH**

Developing new hardware and innovative business models for exploiting access to low Earth orbit.

02.

37 **ALMOST** SPACE FLIGHT

Flights that fall short of reaching into space offer a range of exciting opportunities for both civilian and professional adventurers.

03.

52 **ISLANDS** IN THE SKY

Orbiting stations built by national space agencies are about to be matched by a new generation of platforms for civilian visitors.

04.

75 DESTINATION **MOON**

The Moon is three days' flight time away, and we already know how to reach it. We also know where to build a base once we get there.

05.

99 **INTERPLANETARY** ADVENTURES

Mars has haunted space enthusiasts for generations. Could we send humans to this enticing world, and others besides?

06.

131 ACROSS THE **GULF OF STARS**

Interplanetary space flight is a familiar theme in science fiction. New developments in physics could turn fantasy into fact.

162 credits

164 index

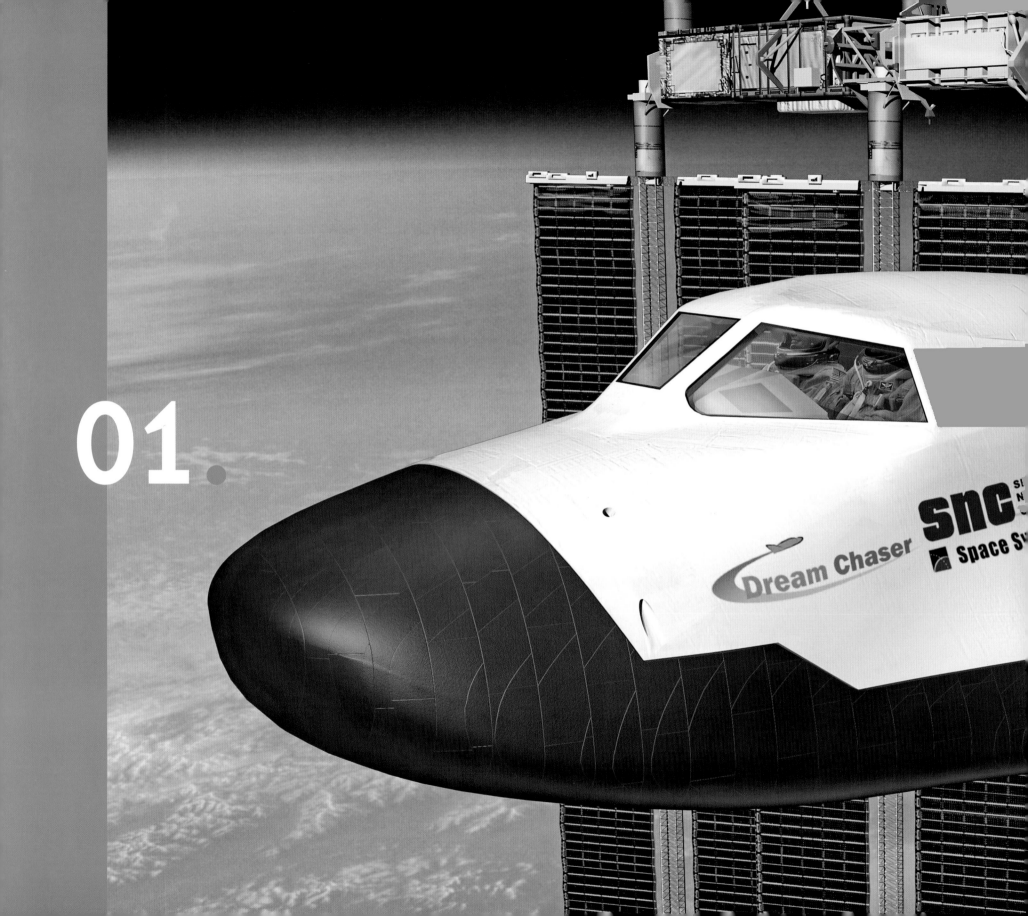

01.

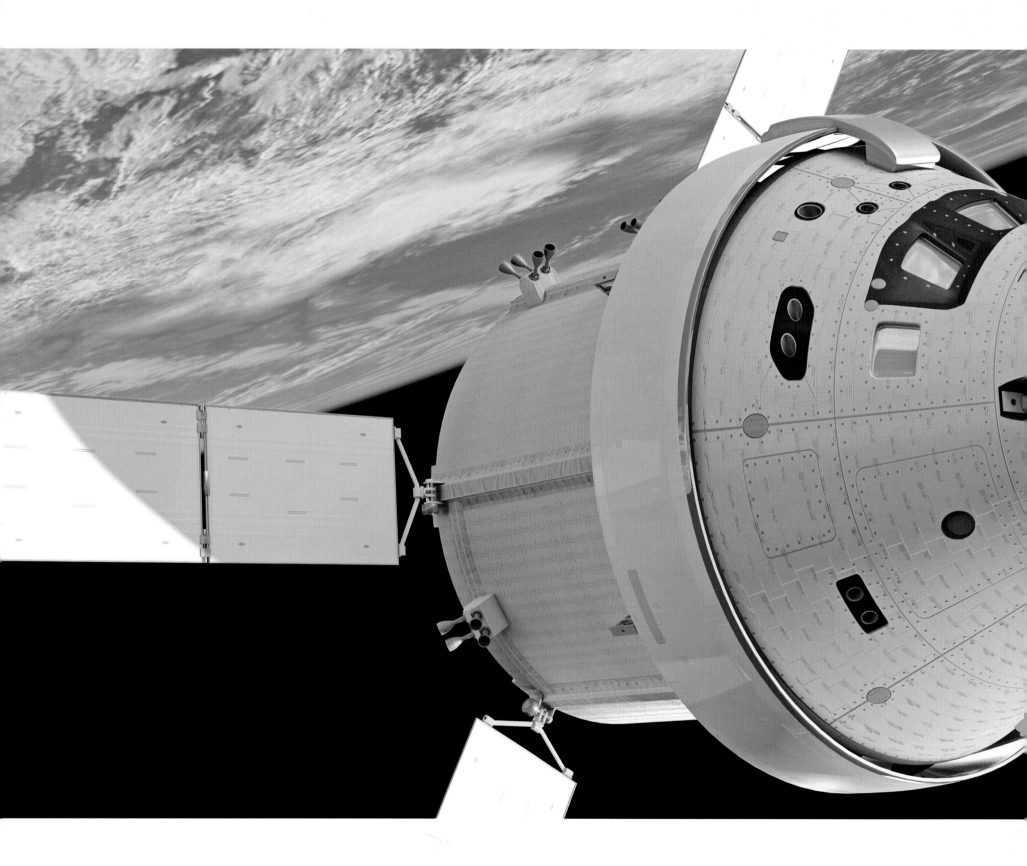

An artist's impression of an *Orion* spacecraft mated to a European-built ATV service module. ⋀

The world's greatest space agency NASA is preparing a vehicle that will extend humanity's reach further into space than ever before. Meanwhile innovative private companies are building new systems for reaching Earth orbit.

ESCAPE **FROM** PLANET EARTH

We are on the verge of a new era in space exploration. For the first time in history, private access to space is becoming almost routine. At this same time, NASA seems to have lost some of its former momentum, waiting for America to make up its mind about the broader future of the national space program. Tough choices lie ahead. The good news is that those choices can be based on a swiftly expanding set of technical possibilities, and mission options.

The journey forward starts on the basis of some difficult backward steps. The space shuttle fleet has retired after thirty years of service. The surviving orbiters, *Atlantis*, *Endeavour*, and *Discovery*, and the glide test prototype, *Enterprise*, are now on display in museums. Our memories of the shuttle system are a mixture of pride and sadness. The first ascent to orbit by *Columbia* in April 1981 thrilled the world, but in January 1986, *Challenger* exploded just seventy-three seconds after liftoff, and all seven crew members were killed. Faulty booster seals and poor management within NASA (the National Aeronautics and Space Administration), the United States' space agency, were blamed for an avoidable disaster. Then the story of space travel and exploration improved, with the repairs

to the Hubble Space Telescope and the assembly of the International Space Station (ISS). Americans once again regarded the shuttle with pride, and perhaps even took it for granted. The shuttle at its best was an adaptable workhorse, an incredible machine capable of transporting astronauts and payloads together.

But as is often the case, pride came before a fall. In February 2003, *Columbia* disintegrated during reentry. Another crew was lost. A suitcase-size piece of thermal insulation foam had peeled off the huge external fuel tank shortly after launch and hit the orbiter's left wing on its leading edge, making a small but ultimately catastrophic hole. Two weeks later, as *Columbia* hurtled through the atmosphere at the end of its mission, hot gases rushed into that hole, destroying internal controls and melting the underlying metal airframe.

President George W. Bush responded to this catastrophe during a televised visit to NASA headquarters in Washington in January 2004. "Today I announce a new plan to explore space and extend a human presence across our solar system," he said. The first goal was to return the remaining shuttle fleet to operational status and finish the construction of ISS. No surprises there, but as his speech

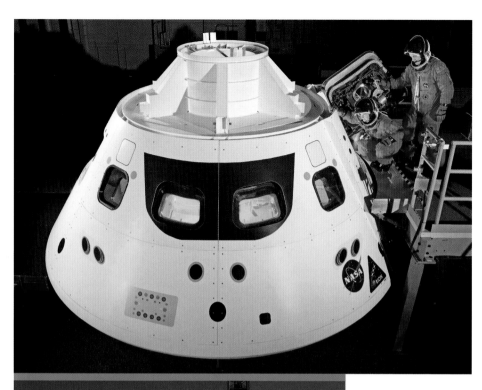

Astronauts prepare to enter an *Orion* engineering mockup to test the capsule's interior layout and seating arrangements. ‹

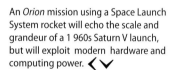

An *Orion* mission using a Space Launch System rocket will echo the scale and grandeur of a 1960s Saturn V launch, but will exploit modern hardware and computing power. ‹ ⌄

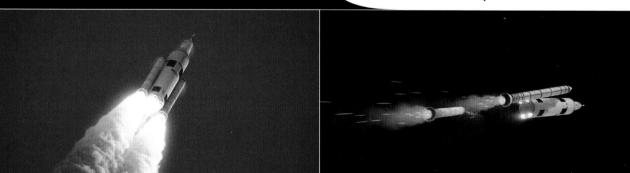

continued, radical new ideas emerged. "Our second goal is to develop and test a new spacecraft, the Crew Exploration Vehicle, by 2008, and to conduct the first manned mission no later than 2014. Our third goal is to return to the Moon. Using the Crew Exploration Vehicle, we will undertake extended human missions to the Moon as early as 2015, with the goal of living and working there. With the experience and knowledge gained on the Moon, we will be ready to take the next steps of space exploration: human missions to Mars and to worlds beyond."

NASA set to work on a new, wingless spacecraft design, with the crew compartment positioned at the top of the launch stack, thereby ensuring that any stray debris peeling away from the flanks of the rocket below would not endanger the crew. The cone-shaped capsule, commonly known as *Orion*, exploits reliable reentry techniques perfected back in the 1960s for Project Apollo. The *Columbia* accident investigators concluded that it is too risky to carry astronauts in the same part of a spacecraft that also contains the propulsion systems, because of the risk of launch failures and the danger of damage to the crew compartment. Investigators noted that the Apollo capsules were remarkably safe. The tough, compact command modules could always be instantly separated from other modules or rockets in the event of failures. When *Apollo 13* suffered an explosion on the way to the Moon in April 1970, the rear service module with the rocket engine was blown wide open, yet the capsule itself was unharmed and returned its crew safely home to Earth.

The space shuttle had no way of separating its crew compartment from the rest of the system. This may always be a problem for any space plane that does not have a separate escape module. NASA has returned to the capsule concept, including an escape rocket that can pull the capsule clear of a wayward launch vehicle. An

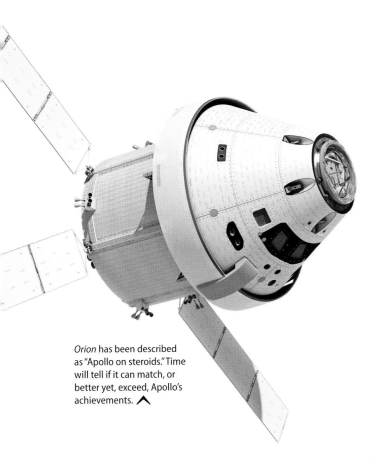

Orion has been described as "Apollo on steroids." Time will tell if it can match, or better yet, exceed, Apollo's achievements. ⋀

improvement in crew safety was just one aspect of a new vision for NASA. *Orion* was incorporated into a grand vision of Moon-Mars exploration, using a fleet of rockets and landers. The program as a whole was called Constellation. Robert Seamans, NASA's deputy administrator in the Apollo era, was one of a dozen enthusiastic veterans called upon to advise *Orion*'s designers. "I served on what they called the 'Greybeard Committee,' all these old hands who knew how we'd reached the Moon first time around. Astronaut John Young was there, and he'd flown in Gemini, plus Apollo, and he'd commanded the first space shuttle flight. We didn't fool around. They put us in a room from eight until five. They brought in food because there was no break for lunch. What we came up with is amazingly similar to what we did with the Apollo." Indeed, NASA officials described their Moon plans as "Apollo on steroids." The proposal, which still exists as a detailed engineering concept, features a lunar landing craft, the *Altair*, carried to orbit unmanned aboard a heavy-lift rocket. Then the *Orion* and its crew locate it in Earth orbit, make a docking and head to the Moon. On arrival, *Altair* detaches from *Orion* and drops toward touchdown. At the end of its surface stay, the lower part of *Altair* stays behind while the upper module blasts back into lunar orbit, makes a rendezvous with *Orion* and transfers its crew. Then *Orion* heads back to Earth for an Apollo-style reentry and splashdown.

There are other significant reasons for reinventing the capsule concept. The shuttles dropped down from low Earth orbit at less than 17,500 miles per hour. *Orion* capsules coming home from deep space missions will slam into the upper atmosphere at nearly 25,000 miles per hour. Winged shuttles cannot sustain that kind of shock. *Orion* uses a blunt and rounded heat shield that is more than capable of surviving high reentry speeds. Why not simply slow an *Orion* as it approaches Earth? Fuel for a powerful

braking burn would have to be carried all the way into deep space and back again, thereby incurring a weight penalty. Most compelling of all, no one would wish to see an engine failure endanger a crew in the last hour of a mission. An *Orion* can hit the atmosphere without having to fire any rocket engines for a braking burn. This is a valuable safety feature.

Development of flightworthy *Orion* systems has continued more or less smoothly ever since President Bush first lent his support to the vehicle. However, even before the banking crisis and credit crunch in the first decade of the 2000s, NASA's plan for a return to the Moon looked ambitious. A year or so after Barack Obama entered the White House in 2009, his administration proposed resetting NASA's priorities yet again: skipping the Moon, putting Mars on the back burner, and aiming instead for an asteroid rendezvous by the mid-2020s. Predictably in hard economic times, NASA had to scale back its ambitions in response to budget constraints. The rocket architecture was adjusted, although the capsule design stayed essentially the same. After all, *Orion* was conceived from the outset as a multipurpose vehicle. As a consequence, design and construction work on this component has proceeded more or less steadily. The capsule taking shape today is America's new national crewed spacecraft. How deep into space it will eventually travel has yet to be determined. A giant new rocket for *Orion* is under development: the Space Launch System (SLS). This will be similar in scale and power to the Saturn Vs that propelled Apollo to the Moon.

Virgin Galactic

There is nothing new about the idea of flying civilians into space. In the late 1960s the now-defunct Pan Am airline held out the promise of routine access to orbit, while bold plans for a resort on the Moon were presented by

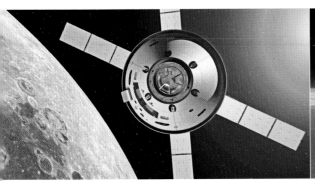

Engineers at the Lockheed Martin company prepare the first *Orion* for flight. ∧

Orion's instrument panel uses flat screen displays to simplify the layout. ⟨

The Space Launch System (SLS) will be assisted at lift-off by solid rocket boosters developed from space shuttle technology. ◄

EUROPEAN SUPPORT FOR ORION

New economic realities have stimulated fresh ideas about international cooperation in space, building on the diplomatic and cultural legacies of ISS. The Orion capsule will be supported by a service and propulsion module that—in all but a few interface details—has already proved its capabilities in flight. The 20-ton Automated Transfer Vehicle (ATV) is the most complex spacecraft ever developed in Europe. Until recently, its main task was to deliver eight tons of crew supplies, propellant and scientific equipment to ISS at intervals of approximately 15 months. The ATV had three times the payload capability of its Russian counterpart, the Progress cargo vehicle.

Although no one was launched aboard an ATV, astronauts wearing regular clothing could board once it was docked. Each ATV became an integral part of ISS for up to six months. During that time, an ATV's engines could be used to reboost ISS's orbit, compensating for atmospheric drag. Technology derived from this hardware is about to have a new lease of life. NASA's Orion spacecraft will be attached to an ATV service module, as depicted in this illustration.

Virgin Galactic's *Enterprise* rocket plane is carried aloft by VMS *Eve*, prior to a rocket-powered test. ◀

SpaceShipOne, under the *White Knight* carrier plane, takes off for its prize-winning mission on October 4, 2004. ▲

the giant Hilton hotel chain. Both these famous brands featured in Stanley Kubrick's highly influential 1968 movie *2001: A Space Odyssey*. The realities of the NASA space shuttle dampened some of these dreams. Despite its many extraordinary achievements, it was not a pathfinder for routine commercial access to orbit, let alone to the Moon. It flew relatively infrequently. It was costly to operate, and its occupants were almost exclusively professional astronauts who knew the risks they were taking on. Even so, in 1994 the American Society of Civil Engineers held a conference in Albuquerque, New Mexico, to explore the possibility of holidays in space. Beautiful models of orbiting hotels and passenger shuttles were presented. Although many delegates were impressed by the perfectly feasible ideas on display, just as many were persuaded that they were impossible dreams. Financial experts warned that ordinary

people would not be willing to pay the extraordinary amounts of cash required to book a space flight.

The pessimists spoke too soon. There are indeed people willing to pay for a taste of space. In 1998, the Virginia-based company Space Adventures negotiated with Russia's Star City cosmonaut training complex to provide private clients with $10,000 weekends of dressing up in space suits and clambering into simulator capsules. Customers also signed up for zero-gravity training flights aboard "vomit comets," planes flying a series of parabolic arcs, enabling passengers to experience a few minutes of weightlessness at the top of each arc.

This was just the more affordable end of the Space Adventures menu. On April 28, 2001, American investment financier (and ex-NASA engineer) Dennis Tito was sealed inside a Soyuz capsule, accompanied by two Russian

Coming home in a Soyuz can be almost as dramatic as the actual space mission. ⋀

Soyuz capsules touch down on land for ease of recovery. Small braking rockets soften an otherwise very bumpy landing. ⋀

RUSSIA'S WORKHORSE

Soyuz's first flight on April 23, 1967, was a disaster. Solo cosmonaut Vladimir Komarov was killed when the landing parachutes failed and his capsule smashed into the ground like an unrestrained meteorite. Subsequent flights went more smoothly, until an even worse disaster on June 29, 1971. Georgy Dobrovolsky, Vladislav Volkov and Viktor Patsayev suffocated when the air leaked from their cabin as they prepared for reentry after a trip to Russia's first space station, Salyut 1. The Soyuz's record since then has been impressive, and although the electronics have been updated, the basic exterior design has hardly changed in five decades. The crewed variant of the Soyuz has flown more than a hundred missions with no fatalities. There have been a few hair-raising close calls, including a launchpad abort when the carrier rocket exploded. Escape rockets pulled the cosmonauts safely away from disaster, and all three lived to tell the tale.

In the absence of the space shuttle, American and European astronauts fly to ISS in Soyuz capsules, at a cost of around $70 million per seat. The cramped capsule cannot solve all future astronaut transport problems. It carries just three people, and two of them must be Russian pilots. The craft is not reusable, the landings are bumpy, and there is minimal control once the crew compartment hits the atmosphere, so the capsules have to descend over remote, unpopulated areas in Kazakhstan, not far from where they are launched in the first place. Soyuz's builders at the RSC-Energia Corporation are anxious to move major launch and recovery operations onto Russian soil, and to update the Soyuz concept altogether. A replacement vehicle, offering more crew space, and greater aerodynamic control during the reentry phase, may become operational for the year 2020.

At least two Soyuz vehicles docked to the ISS serve as crew transport systems, doubling as emergency lifeboats in case of trouble. ⟨

cosmonauts on their way for a tour of duty aboard the ISS. Tito had committed $20 million of his own cash to secure this opportunity. He booked his flight through Space Adventures, with cooperation from the Russian Federal Space Agency (commonly called Rocosmos). At first, NASA was not keen on his mission. Tito had to remain inside the Russian modules of ISS throughout his week's sojourn. But his Russian hosts made him feel welcome. Cosmonaut Yuri Baturin told him, "We are very happy to accompany you to space. We like your mathematical mind, and we like even more your romantic soul."

A year after Tito's flight, South African Internet entrepreneur Mark Shuttleworth made a similar voyage, and NASA's rules concerning access to other areas of ISS were relaxed. On his return to Earth, Shuttleworth set up a foundation to encourage the teaching of math and science in Africa. So far, space tourists have shown as much dedication to their missions as any professional astronaut. In October 2005, when Greg Olsen became the third private entrepreneur to visit the station, he was mildly disturbed to be thought of as a tourist. "I spent over nine hundred hours in training. There were many exams, medical and physical, as well as classroom and competency tests. It's not like you pay your money and go on a ride. You have to qualify for this." As the founder of a New Jersey company specializing in optical and infrared sensors, he conducted serious scientific experiments during his time in orbit. He took care, however, not to compare his qualities with those of long-term professionals. "I'm a space traveler, and I have been in orbit, but I have far too much respect for astronauts and cosmonauts to call myself that."

Mark Shuttleworth training for his Soyuz flight to ISS. ∧

One private citizen has funded his way to space not once, but twice. Hungarian-American software entrepreneur Charles Simonyi, the man who oversaw development of Microsoft's Office suite of applications, flew to ISS in April 2007 and again in March 2009. While entrepreneurs of such prominence and determination may have no great difficulty raising $20 million in travel expenses, we cannot expect too many repeat performances at those prices. The future of personal space flight will rely on reducing ticket costs to some thousands, rather than several millions, of dollars. In 2002, market researchers Zogby International surveyed over four hundred American business leaders and wealth-creating individuals about their appetite for space. One in five confirmed they would be happy to pay up to $250,000 for a suborbital flight, while just seven in a hundred was willing to contemplate $20 million for a sojourn in orbit. A happy balance of price points could create a vibrant market.

Getting Us There

Access to space for private citizens will develop once a genuinely cost-efficient spaceliner becomes available. In May 1996 a group of businesspeople held a gala dinner in St. Louis, Missouri, to celebrate Charles Lindbergh's historic 1927 first flight across the Atlantic Ocean, funded by an earlier generation of patrons from this Midwest city. The patrons had been competing for a $25,000 prize set up by New York hotelier Raymond Orteig, which was a substantial sum of money in 1927. Could a similar prize spur modern industry to develop cheap, affordable human space flight? And so the $10 million Ansari X Prize was created, under the leadership of Space Adventures cofounder Peter Diamandis. Its aim: to reward the first private reusable passenger-carrying ship to reach the edge of space. Diamandis's partner in this prize initiative was a remarkable woman called Anousheh Ansari. She spoke no English when she emmigrated to the United States from Iran in 1984 at the age of sixteen, but she hoped that living in the United States would help her realize her dream of becoming an astronaut. She enrolled at George Mason University, outside of Washington, D.C., to study electrical engineering. There she met Amir Ansari and his brother Hamid, the latter of whom she married in 1991. The three of them founded Telecom Technologies Incorporated. By 1996 it was the

Russia's venerable Soyuz design is showing its age, as this view of the distinctly cramped interior demonstrates. ⟨

A recovery helicopter captures a dramatic night time Soyuz recovery operation. ⟨⟨

fifth-fastest-growing technology company in Dallas, Texas. Anousheh never lost her fascination for the human adventure of space flight, and she and her family were major sponsors of the Ansari X Prize. "As a child I looked at the stars and dreamed of being able to travel into space," she said when her financial involvement was announced. "As an adult, I understand that the only way this dream will become a reality is with the participation of private industry and the creative passion of smart entrepreneurs."

A dozen companies took up Ansari's challenge. A fascinating array of designs included delta-winged space planes, capsules shaped like seedpods, and one machine known as *Roton*, which featured rocket-powered helicopter blades. Some competitors built real hardware, while others were unable to finance much beyond computer-generated concept artworks or unpowered mockups. But the seeds for a new industry were sown, and a number of companies that did not make the cut first time around are still in business. They are older, wiser, and less naive about the financial burdens of research and development in a complex field.

As a young man, pioneering aircraft designer Burt Rutan was inspired by NASA's quest for the Moon and its many other bold and fast-paced achievements throughout the 1960s. He looked forward to similarly exciting new developments beyond Project Apollo. The decades dragged on, and the space agency became somewhat slower and more cautious in its ambitions, which were constrained mainly by shrinking budgets and escalating costs. Rutan grew impatient and decided to shoot for the X Prize by creating the first purely privately funded human-carrying space vehicle. "Government space agencies want to commit us to their old-fashioned technologies," he announced. "We already know how that stuff works. What we need is the freedom to try some new, smarter, and less expensive ideas."

The company that Rutan founded, Scaled Composites in Mojave, California, is renowned for creating lightweight, fuel-efficient aircraft of exceptional beauty and elegance. Its most startling creation, *SpaceShipOne*, claimed the X Prize on October 4, 2004, when pilot Brian Binnie took the craft to an altitude of 69.5 miles. The region of sky where the Earth's atmosphere ends and space begins is nebulous at best. International convention defines the boundary as 100 kilometers, or 62 miles. *SpaceShipOne* ascended well past that boundary.

The project immediately earned back half the $20

X Prize co-founder Anousheh Ansari on board a Soyuz as it heads toward the International Space Station. Ansari is wearing a Russian Sokol space suit. ❯

The *Lynx* is neat and compact. It requires no additional carrier craft. ⌄

First-generation *Lynx* vehicles will not fly all the way into space, but the altitude difference will only be of concern to pedants. ⟨

THRILL-RIDE TO THE EDGE OF SPACE

XCOR Aerospace was founded in 1999 by veterans of the Roton helicopter-rocket project. The new company specializes in fitting small rocket motors to the rear of small aircraft, literally boosting their ability to climb into the sky. Rocket plane racing may soon be as common as conventional air races and acrobatics. XCOR's flagship project is the Lynx, a plane that uses rocket power to reach an altitude of around 200,000 feet: about two thirds the way to space. One pilot is accompanied by one fee-paying client.

million invested in the vehicle by Microsoft cofounder Paul Allen. Just as Charles Lindbergh's flight inspired a new transatlantic air industry, Rutan's team generated a similar momentum. On the ground, watching *SpaceShipOne*'s prize-winning smoke trail through powerful binoculars, were two lifelong space fans from Britain: the entrepreneur and businessman, Richard Branson, and his colleague Will Whitehorn. Two days later, Branson announced that his Virgin investment group was ready to finance *SpaceShipOne*'s larger successor, along with supporting ground facilities at a dedicated site in New Mexico. An additional $100 million was pledged for building a small fleet of Virgin Galactic suborbital spaceliners, each capable of lifting six passengers and two pilots. "Someday, children around the world will wonder why we ever thought space travel was just a dream we read about in books or watched, with longing, in Hollywood movies," Branson explained while announcing his venture to a startled world. "If we can make space fun, the rest will follow. This is a business that has no limits." Hundreds of customers paid substantial deposits against their $200,000 ticket fees, essentially making them partial funders of the new spacecraft's development and construction.

For the next four years, Virgin Galactic and its collaborators at Scaled Composites maintained a relatively discreet profile. At last, the hangar doors were opened to reveal Virgin Mothership (VMS) *Eve*, described by Branson as "one of the most beautiful and extraordinary aviation vehicles ever developed." *Eve* is a twin-fuselage jet aircraft capable of lifting the passenger-carrying Virgin Spaceship (VSS) *Enterprise* to the uppermost levels of Earth's atmosphere and releasing it for a blast into space. Twice the size of the *White Knight* carrier plane that lifted *SpaceShipOne* into the air in 2004, *Eve* is powered by four Pratt and Whitney jet engines, but the real work of lifting this aircraft into the sky is conducted by its wing, a continuous strip of reinforced carbon composite materials 140 feet long: the largest of its kind ever constructed. The smaller, winged rocket plane *Enterprise* was unveiled eighteen months after *Eve*'s first public appearance.

Enterprise is suspended beneath *Eve*'s middle wing section and carried to a launch altitude of ten miles. *Enterprise* then drops away, ignites its rocket motor, and climbs toward space, accelerating to three times the speed of sound. Meanwhile *Eve* heads back home for a conventional landing. Just two minutes after release,

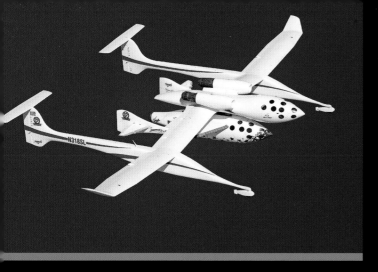

Enterprise is in space, and its occupants are officially astronauts. They experience at least five minutes of weightlessness, drifting free of their couches and staring at the Earth through the vehicle's large round windows, no doubt, taking plenty of pictures.

When rocket plane *Enterprise* begins to fall back toward Earth, it repositions its wings perpendicularly relative to the fuselage and reenters the atmosphere belly first, generating maximum air resistance. At just over thirteen miles above the ground, the wings rotate back to the horizontal position, and the vehicle glides home to a runway landing, ready to be refueled and flown again within a few days. *Eve* and *Enterprise* are forerunners of a fleet of vehicles being prepared for commercial flight.

Good for the Environment?

Former Virgin Galactic chief executive Whitehorn, the man who kick-started the project alongside his close friend Branson, maintains close links with the company. He is sure that personal space flight is just one of several markets that the VMS and VSS systems can exploit. "The spacecraft takes eight people up, including the two pilots, and brings them all down again to a safe landing. It's a glider, with wings, landing gear, and of course, all the life support for the human occupants," he explains. "Now imagine if we didn't have the people, and we didn't have to bring any of the machine down to Earth again. Instead, you have a slender expendable rocket, tipped by a satellite payload. VMS could launch small satellites all the way into full orbital space." An unmanned rocket, LauncherOne, slung under a VMS, will deliver 500 pounds of payload to low Earth orbit and at least 200 pounds to higher altitudes. Virgin Galactic's other hope for its new technology is that it might one day help reduce the aviation industry's carbon footprint. If *Eve*, *Enterprise*, and their sister ships can do the hard part—getting people into suborbital space—using composite materials instead of metal structures, then it should be possible for conventional airliners to reduce thei dependence on aluminum and titanium for their wings and fuselages. Composites are lighter than metals and are therefore much more fuel-efficient. Virgin Galactic's technology might actually benefit the environment, just so long as everything works according to plan. As Whitehorn points out, "Aviation is being unfairly picked on by the

SWINGING THE WINGS

Ships returning from a 17,500-mile-per-hour orbit have little choice but to meet the reentry challenge head-on, but a suborbital craft has a more gentle option. It can minimize its reentry problems by losing most of its speed on the way up. This might sound crazy, given all the fuel and energy expended on the rocket-powered final ascent, but the object is not to stay in space for more than a few minutes. Think of a tennis ball thrown high into the air. At the top of its arc it hovers for the briefest moment, caught precisely between the force of gravity pulling it downwards and the last dregs of upward momentum imparted by your throwing arm. And then gravity wins completely and the ball falls back down to Earth. The key point is that although you threw that ball upwards as hard as you could, it slows almost to a standstill at the highest point of its trajectory.

Burt Rutan was inspired by the example of a shuttlecock hurled aloft by a badminton racquet stroke. The shuttlecock runs out of momentum at the top of its arc, then drifts back to Earth, slowed by the aerodynamic drag of its feathers. In fact the repositioning of the spaceship's wings during descent is specifically called "feathering." This technique has been scaled up for Virgin Galactic's VSS *Enterprise*.

Burt Rutan demonstrates the unique "feathering" wings created for a new generation of suborbital rocket planes. ▲

NEW DEPARTURES?

SpacePort America is located in the southern portion of Sierra County, 45 miles north of Las Cruces, New Mexico. In April 2008, county voters agreed to a modest increase in a local sales tax, to assist construction of the Spaceport, thereby boosting economic activity and local employment. Sierra County joined adjacent Dona Ana County in forming a special Tax Increment Development District. The business of private space flight is now embedded into the everyday economy. Virgin Galactic is the most prominent user of the new complex, but will not be the only company basing is operations here.

New Mexico boasts calm, bright skies throughout most of the year. Its sparse population is at low risk from any wayward spacecraft, while the airspace above the launch site is already clear of conventional airliner traffic because of the nearby White Sands missile testing range. All being well, money will flow into the project: not just from privileged suborbital passengers, but also from visitors and tourists making day trips to Spaceport America, simply to watch the various vehicles in flight. Many will need overnight accommodation in nearby communities. They will also expect to purchase meals and soft drinks, souvenirs, books, toys, postcards, and so on. Most jobs around Spaceport America will be only peripherally to do with actually flying or servicing spacecraft.

From the air, the terminal building and hangar complex looks like a cross between a stingray and a flying saucer, with a Star Wars-style suggestion of the Millennium Falcon's crab-like shape. Renowned British architect Norman Foster and his American partners at U.R.S. Corporation have delivered a practical and eco-friendly design that fits elegantly into the natural landscape. Certain vantage points make the hangar seem like nothing more than a gentle mound rising softly from the terrain as though it has always been there.

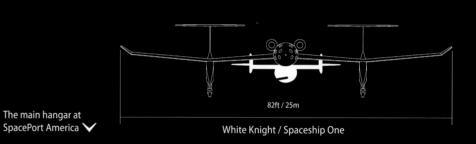

The main hangar at SpacePort America ⌄

82ft / 25m

White Knight / Spaceship One

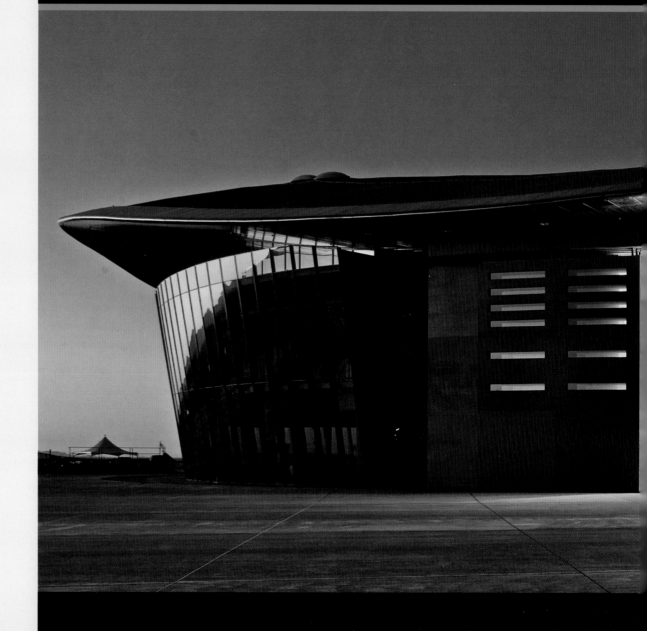

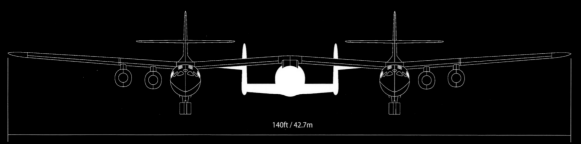

140ft / 42.7m

Virgin Mothership / Spaceship Two

"THIS TYPE OF PROJECT NEEDS PEOPLE WITH FIRE IN THEIR EYES"

Burt Rutan, 2006

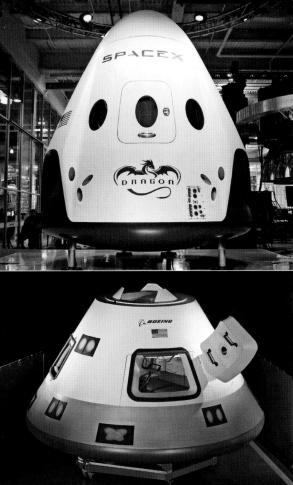

Dragon technology (top) may face competition from Boeing's proposal, the CST-100 (bottom). ◀

is based in Hawthorne, California. The company was established in 2002 by Elon Musk, the man who built PayPal into an essential Internet tool before selling it to eBay. Musk risked $100 million of his own cash to start SpaceX while soliciting further investment, including from a NASA program known as Commercial Orbital Transportation Services (COTS). Under this program, private hardware manufacturers are awarded staged contracts on achieving particular milestones. The emphasis is on providing launch services rather than adherence to specific space agency designs. Under COTS, manufacturers can choose the configuration of their vehicles, so long as they deliver the services that NASA needs while satisfying its stringent safety requirements.

Another advantage of this kind of funding agreement is that SpaceX and other companies involved in COTS are free to sell launch services to clients apart from NASA, because the space agency essentially buys the ride, not the horse. The very nature of a government space agency makes its operations expensive. For the sake of fairness, any tax-funded space program must request bids from a range of hardware manufacturers, whether of ground support systems or entire spacecraft. Even when agency chiefs suspect that a particular bid will not be appropriate, time and resources must be allocated to a transparent selection process. As the great rocket scientist Wernher von Braun once said, "We can get to the Moon in ten years, but the paperwork will take longer." By way of contrast, smaller, leaner, and purely private companies such as SpaceX can operate more efficiently by working directly with favored subcontractors. Simplifying the bureaucracy can speed up development while keeping costs down. This is not to suggest that private enterprise is better than NASA's way of doing things. The U.S. Congress can raise far more cash for spaceships than any private company can. Even so, a distinct mood of change is in the air.

SpaceX's latest rocket, *Falcon 9*, is so named because of its cluster of nine Merlin engines: an array powerful enough to lift the company's flagship *Dragon* capsule, whose primary task is hauling cargo into orbit. This is just the prelude to a human-rated system. *Dragon* is a pressurized vehicle that docks with ISS and returns safely to Earth. In theory, someone stowing aboard would reach space and come home again unharmed. In practice, an integral launch escape system and life support equipment will be needed before *Dragon* can ferry crews. These are under development. The crew-carrying variant, called

green lobby. There are half a billion computer servers in the world, all dependent on carbon-derived energy when they are manufactured, and all of them constantly soaking up energy and pumping out heat during use. The endless growth of the Internet has overtaken aviation in terms of its carbon dioxide output."

Falcons and Dragons

Virgin Galactic's rocket plane is one of the most impressive developments in aerospace history. Even so, its mission above the atmosphere is brief, and it cannot reach full orbit, let alone make a rendezvous with ISS, which operates at 230 miles above the ground and arcs across the heavens at more than 17,000 miles per hour. Catching up and docking with ISS takes serious rocket power. The retirement of NASA's shuttle fleet has stimulated a new market for spacecraft capable of servicing ISS's needs.

Space Exploration Technologies Corporation (SpaceX)

SpaceX's proposed design for a future crew capsule accommodates seven astronauts. ◀

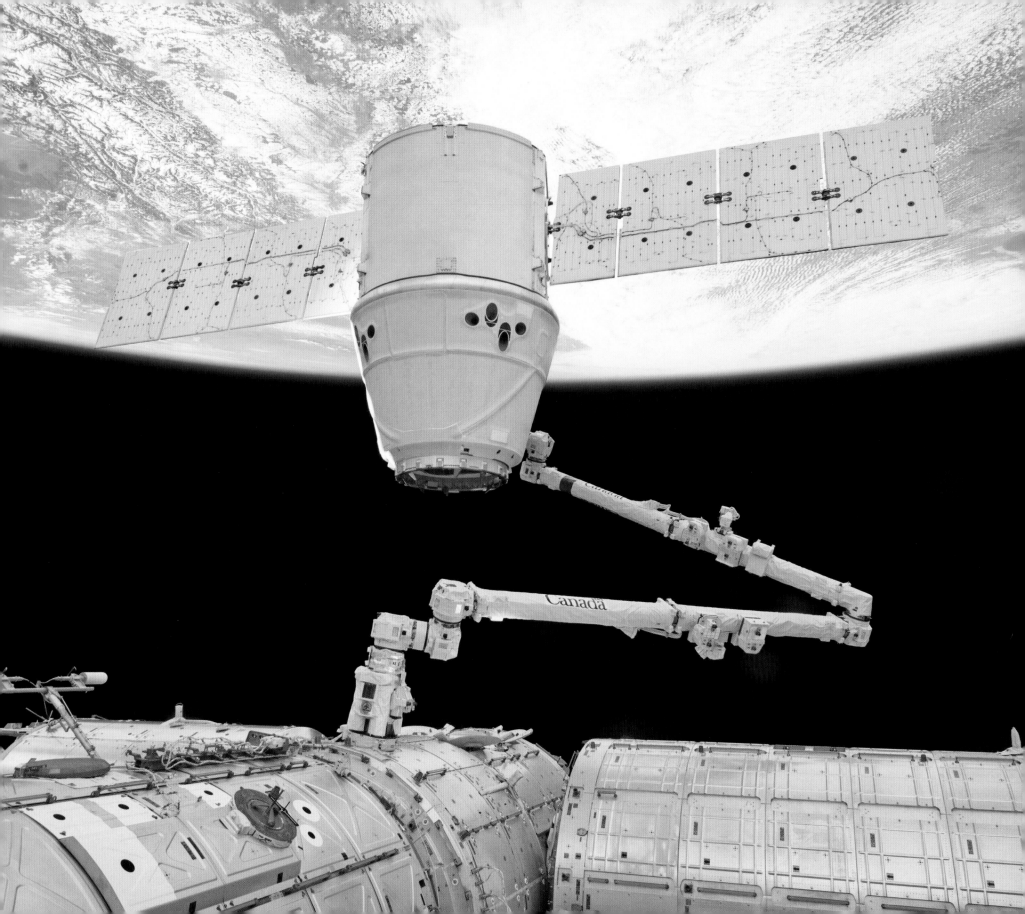

A *Dragon* spacecraft, solar panels extended, becomes an inhabitable extension of the ISS until the time comes for undocking and reentry. ◀

Dragon capsules are recovered at sea. Future variants will be fully reusable. ▼

DRAGONS...

The pressurized section of a Dragon capsule is designed to carry both cargo and humans into space. Toward the base of the capsule, patented Draco rocket thrusters and fuel reserves are accommodated between the outer shell and the inner pressurized compartment. Thrusters for the crewed DragonRider may double as an escape system, powering the capsule clear of a launch vehicle in the event of problems. The cylindrical rear module trunk supports the spacecraft during ascent to space, carries unpressurized cargo, and houses Dragon's solar arrays. The service module remains attached to Dragon until shortly before reentry into Earth's atmosphere, when it is jettisoned. After descending under parachutes, Dragon splashes down in the Pacific. Future re-usable variants may touch down on land, using Draco thrusters to soften the impact.

DragonRider, will carry up to seven astronauts. This will be especially useful if all the occupants aboard ISS need to be evacuated in a hurry. Once docked, the ship's power systems will remain viable for at least six months.

A *DragonRider* capsule and its supporting systems will weigh in at around ten tons when fully fueled for launch. A future generation of carrier vehicle, the *Falcon* Heavy, will lift fifty tons to orbit: equivalent to a fueled Boeing 737 airliner complete with passengers, flight crew, and all their luggage. Musk is certainly ambitious. In April 2011 he announced that this new rocket "will carry more payload to orbit than any vehicle in history, apart from the Saturn V Moon rocket, which was decommissioned after the Apollo program. This opens a new world of capability for both government and commercial space missions." Powered by a cluster of three Falcon 9 first stages, the *Falcon* Heavy should launch twice the payload of a space shuttle at less than one-tenth the cost of a shuttle launch.

Of course history has seen such optimistic projections before from rocket manufacturers, but SpaceX achieved an orbital spacecraft, two generations of launch vehicle, and far more besides, on an expenditure of around $1 billion across its first decade of operations. This is a fraction of the expense traditionally associated with the rocket business. So far, so good. If SpaceX can find clients with sufficiently large payloads to justify the *Falcon* Heavy, the way will be clear for some very ambitious new capabilities in space, as long as the launch market, whether civilian or military, creates sufficient demand. New rockets must compete with existing vehicles. These markets are finite, and someone, somewhere, must eventually fail to win that crucial share of business that makes the difference between survival or shuttering their company. If anyone imagines that today's breed of private space entrepreneurs are just playing with their money, it's worth remembering that by 2008, Musk had used up almost all the cash from the PayPal sale, and freely admitted that SpaceX was "running on fumes" in the wake of three discouraging test launches of the Falcon 1 rocket. It took courage and commitment to keep the company on track. The fourth, flawless launch of a *Falcon 1* in September of that year turned SpaceX's fortunes around, paving the way for development of the much larger and more powerful *Falcon 9* launch vehicle, along with its key payload, *Dragon*.

Competition from relative newcomers is not always

Microsoft co-founder Paul Allen helped fund Burt Rutan's SpaceShipOne. Allen is now promoting a massive air-launch system, called *StratoLaunch*. ⋀

...AND SWANS

Orbital Sciences Corporation, based in Dulles, Virginia, established its reputation with Pegasus, the world's first independently developed space launch vehicle, which made its operational debut in 1990. The slender rocket, approximately the size of a cruise missile, is launched mid-air from under the wing of a conventional jet plane, a Lockheed L-1011 known as Stargazer. Pegasus can deliver one-ton payloads to low Earth orbit, for very modest costs. Today the company's much larger Antares rocket lifts the Cygnus ("Swan") cargo supply vehicle toward a semi-automated rendezvous with ISS. Meanwhile, the Pegasus principle is being upgraded, and on a huge scale. Paul Allen is working with Orbital Sciences, and Scaled Composites, on a massive air-launch system called Stratolaunch.

Orbital Science's air-launched Pegasus was the first privately developed rocket capable of delivering payloads to space ‹

A Cygnus cargo vehicle, built by Orbital Sciences. Future variants may support deep space missions. ›

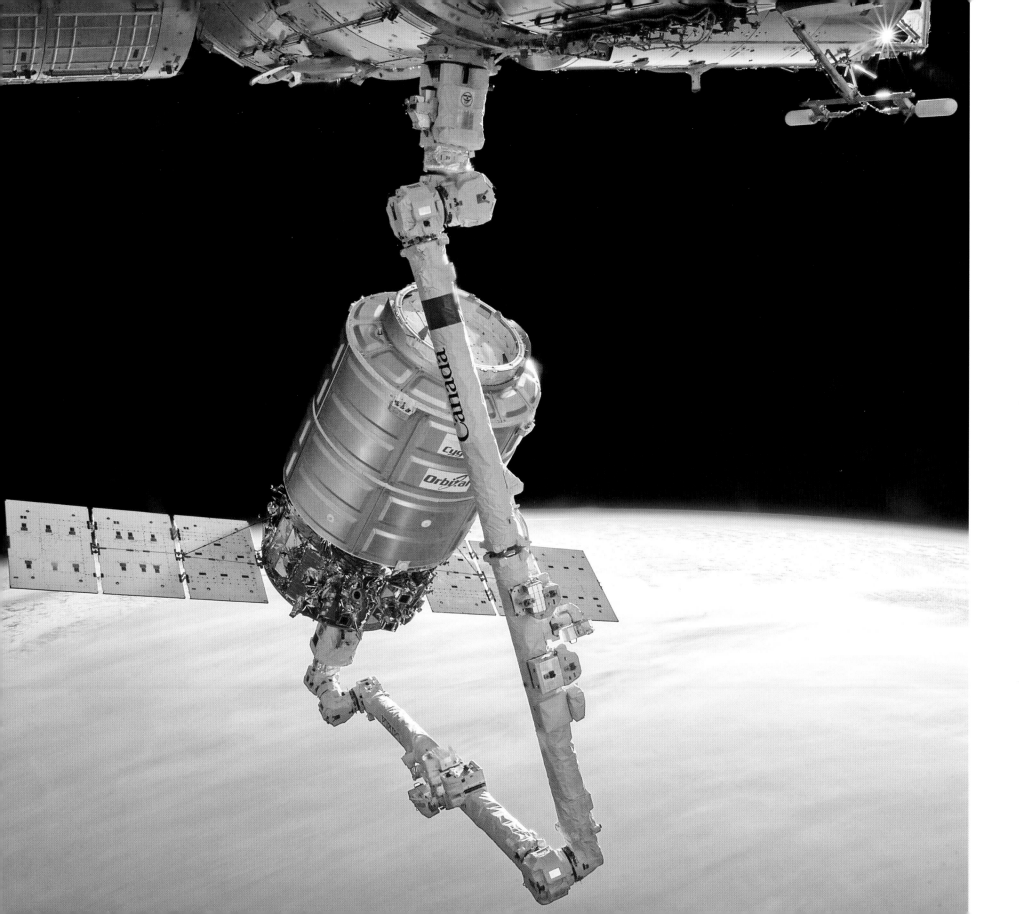

FEET FOR FALCONS

The Moon may be out of reach for *Tintin*-style spaceships, but the dream of Vertical Take-off and Landing, or VTOL, continues to haunt many rocket designers, for the simple reason that everyone is exasperated by the idea of throwing away expensive rocket stages after just one flight. Elon Musk's SpaceX company is already flying some of the hardware associated with a landing system. The plan is for the expended first stages of Falcon rockets to fall back to the Earth under parachutes, then deploy shock absorbing legs as they near the ground. Accurately targeted touchdowns on land, cushioned by retro engines, could preserve precious machinery for re-use. A small proportion of the fuel normally consumed during lift-off and ascent would be held in reserve for the touchdown burn.

Swiss engineers are developing uncrewed lifting body spacecraft for air launch. ⏶

SpaceX is determined to recover spent rocket stages for refurbishment and relaunch. ❯

welcomed by the more traditional, large-scale aerospace manufacturers, some of whom have accumulated more than half a century's worth of experience in building space systems, from Saturn V rockets to ISS modules. In their capacity as major employers, a key handful of corporations exercise a certain degree of political influence over national space policy. Little wonder that COTS contenders tend to win only a modest share of NASA's budget. The most efficient company is not necessarily the one that creates the greatest number of jobs. Lawmakers in Congress bring to the table their own agendas when they debate national space policies. Some favor swift, efficient access to space, while others believe it is their proper and legitimate duty to protect jobs on the ground. Closing down giant aerospace plants in favor of leaner enterprises is not a pleasant option for anyone in the corridors of power. Then again, there aren't many members of Congress who relish massive space spending, either. NASA's plans evolve always into a balance between engineering truths and political

private space tourists

DECEMBER 1990
Japanese TV journalist **Toyohiro Akayima** accompanies a Russian crew to the Mir station. His flight is funded by the Tokyo Broadcasting System.

MAY 1991
British food chemist **Helen Sharman** is selected from 13,000 applicants and flies to Mir, funded by the Russian government after a sponsorship scheme fell short of its targets.

MAY 2001
Dennis Tito becomes the first human to finance his own trip into space, spending a week on board the Russian modules of the International Space Station (ISS) at a cost of $20 million.

MAY 2002
South African internet entrepreneur **Mark Shuttleworth** is granted access to all areas of ISS during his sojourn.

JUNE 2004
Burt Rutan's SpaceShipOne, piloted by **Brian Binnie**, becomes the first privately funded human-carrying space vehicle to leave the Earth's atmosphere.

Technologies for air launch are already available. All that's needed is funding. ⋀

OCTOBER 2005
American industrialist **Gregory Olsen** becomes the third individual personally funding a trip to orbit and a stay aboard ISS.

SEPTEMBER 2006
Iranian-born **Anousheh Ansari**, co-founder of the Ansari X Prize, is the first woman to finance a trip to space, staying aboard ISS.

APRIL 2007
Charles Simonyi, the Hungarian-American creator of Microsoft Office, reaches ISS for a two-week stay in orbit.

OCTOBER 2008
Computer games developer **Richard Garriott**, son of NASA Skylab astronaut Owen Garriott, fulfills a lifetime ambition and spends 12 days in orbit.

MARCH 2009
Another significant milestone for space flight is logged, as **Charles Simonyi** becomes the first person to make a second privately funded trip to orbit.

SEPTEMBER 2009
Canadian performance artist **Guy Laliberté** brings a fresh creative mind to space flight, using his stay aboard ISS to promote environmental awareness.

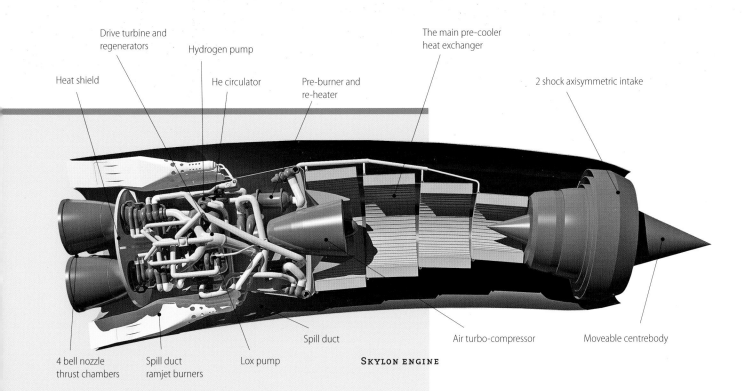

Drive turbine and regenerators

Hydrogen pump

He circulator

Heat shield

Pre-burner and re-heater

The main pre-cooler heat exchanger

2 shock axisymmetric intake

Spill duct

Air turbo-compressor

Moveable centrebody

SKYLON ENGINE

4 bell nozzle thrust chambers

Spill duct ramjet burners

Lox pump

A BRITISH TRIUMPH?

A British team of engineers, led by Alan Bond, is creating a new engine that could revolutionize access to space, blurring forever the distinction between aircraft and spacecraft. The Synergetic Air-Breathing Rocket Engine, or SABRE, exploits air while in atmospheric flight, just like a jet engine, then switches to an on-board oxygen supply while in space, just like a rocket engine. Bond's Skylon space plane has been under development for several decades. Recent successes with the SABRE development have spurred significant international space agency interest in these technologies. Here we see a Skylon docked to a space station built from Bigelow Aerospace's inflatable modules.

A Skylon docked to a commercial space station featuring inflatable Bigelow modules. ◄

1 Ceramic Aeroshell
2 Canards
3 Liquid Hydrogen tanks
4 Liquid Oxygen tanks
5 Payload Bay
6 Avionics
7 Air Intake
8 Heat Exchangers
9 SABRE Engine
10 Orgital Thrusters

SKYLON CUTAWAY

necessities. Within this restless environment, a persistent new breed of companies may yet determine the future of space flight.

Space Planes or Capsules?

The outer shell of an Earth return spacecraft is heavily shielded against the heat of reentry. Just as important is the shock wave, or bow shock, of tightly compressed air that the craft creates just ahead of it. The streamlined shape of a typical airplane is designed to reduce atmospheric drag by minimizing bow shocks. This is why most jets and airliners look so sleek. However, at reentry speeds of several thousands of miles per hour, a spacecraft's bow shock needs to be deliberately obstructive so that it slows down the vehicle rather than easing its passage. The shocked layers of air also help insulate the skin of the spacecraft against the friction of the atmosphere.

Slender wings on a spacecraft are great for making a controlled landing on a runway at the end of a mission, but they can be a nuisance if their bow shocks are so thin that they fail to create a decent insulation layer. Conventionally shaped wings on a vehicle descending from full orbital velocity would simply burn away. Another problem is one of fuel economy. In the vacuum of space, wings are dead weight, because they have no job to do; yet the launch system has to expend fuel carrying their mass into orbit.

"FIRST COMES THE IDEA, THE FANTASY. THEN COMES THE REALITY"

Russian theorist Konstantin Tsiolkovsky, 1926

SHUTTLE REVIVAL

Sierra Nevada's, a partially winged lifting body, draws on the HL-20 mini shuttle project, an abandoned but technically well reasoned NASA proposal from the late 1980s. Dream Chaser is funded in part by NASA's Commercial Crew program, the logical follow-on from the cargo-based COTS scheme. Sierra Nevada has other clients in mind, too. European space agency ESA has taken an active interest in DreamChaser, as all parties in ISS look to the future possibilities of crew access. NASA would prefer to devote its Orion spacecraft to deep space flights, while leaving low Earth orbital business to new private vehicles.

Sierra Nevada's *Dream Chaser* spacecraft poses for a beauty shot at NASA's Dryden Center in California. ▼

NASA has worked extensively with "lifting body" vehicles, in which the distinction between winged and capsule configurations is deliberately blurred. Lifting bodies can be steered during atmospheric flight, like aircraft, but at the same time they create protective bow shocks, like capsules. A lifting body's contoured body shape provides winglike aerodynamic control, yet is sufficiently fat that the interior volume can accommodate useful mass, such as life support, fuel tanks, or other payloads. Experiments throughout the 1960s preceded the development of the space shuttle, whose thick wings had softly rounded leading edges. The space shuttles were never capable of deep space missions, whose returning crew modules must survive a 25,000 miles-per-hour reentry. The *Orion* capsule will be better suited for that challenge. For low Earth orbit operations, a new generation of lifting body–style minishuttles may yet be valuable because they can land with pinpoint accuracy. It is difficult and expensive to pluck capsules out of the ocean.

China in Space

For the best part of half a century, only two nations possessed the ability to launch humans into space. Russia began the adventure in April 1961 by launching cosmonaut Yuri Gagarin into orbit. Three weeks later, America launched Alan Shepard on a fifteen-minute suborbital arc, riding a cramped Mercury capsule atop a converted Redstone ballistic missile. The Space Race had begun. On July 21, 1969, America became the winner when Neil Armstrong stepped onto the Moon. Russia abandoned the Moon, concentrating instead on Earth orbit.

But there was another player waiting in the wings. China pursued the development a manned spacecraft in the late 1960s, until the project was cancelled in 1972. There seemed no obvious reason to send Chinese astronauts into space after the Apollo lunar landings had been accomplished; and anyway, the Chinese economy at that time was not up to such a challenge. The mood in modern Beijing, however, has shifted dramatically. Arguments with America over the fate of Taiwan and disagreements

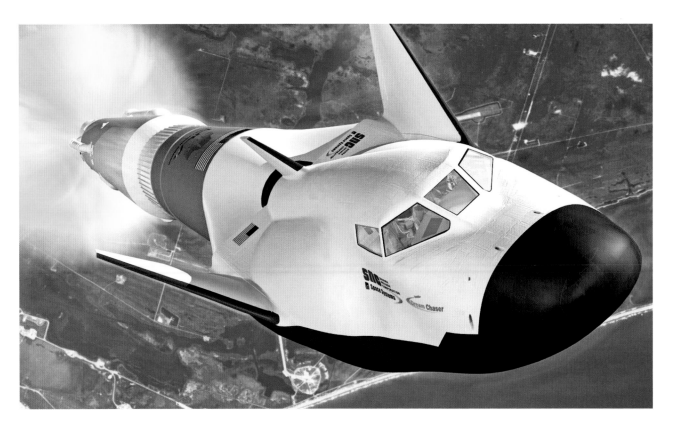

Dream Chaser will be launched by an Atlas-V rocket. ❯

02.

The higher reaches of our atmosphere offer new opportunities for adventure, whether at rocket-driven hypersonic speeds, or by ascending at just a few tens of feet per minute, using almost no fuel.

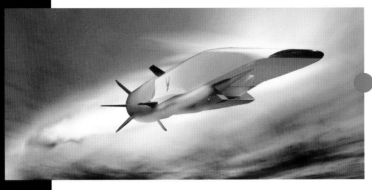

ALMOST
SPACE FLIGHT

Waverider proved that air-breathing scramjets can reach 6,000 miles per hour. ︿

Lockheed Martin hopes to build SR-72, an uncrewed hypersonic drone for reconnaisance missions. ❮

Jet and rocket engines are similar beasts. A mix of fuel and oxidizer is forced into a combustion chamber at high pressure. Then the mix is ignited. The resulting exhaust plume, expelled from the rear of the engine, propels the surrounding vehicle. In a rocket engine, that essential mix is fed into the combustion chamber by turbo pumps linked to supply tanks within the vehicle. A jet engine exploits oxygen freely available from the surrounding atmosphere, scooping it through prominent air intakes at the front and using compressor fan blades to pressurize the air and force it into the combustion chamber. Jets, of course, run out of useful atmosphere at around fifty thousand feet. However, an extremely fast jet, known as a scramjet, can continue to exploit the thin upper atmosphere simply by virtue of scooping it up so fast that the tiny traces of available oxygen accumulate rapidly enough to keep the engine firing.

Scramjet technology has already been tested by NASA. In November 2004 an unmanned X-43 craft was accelerated to nine times the speed of sound. Three years later, Australia's Defense Science and Technology Organization (DSTO) flew its equivalent machine to an altitude of 330 miles, where it flew ten times faster than sound: over six thousand miles per hour. Boeing's unmanned X-51 Waverider achieved similar results in 2010. One drawback is that a scramjet vehicle has to be moving extremely fast before its engine can be lit. A rocket "kicker" stage is usually needed to push the vehicle toward cruising speed. The X-43, for instance, was carried aloft under the wing of a NASA B-52 aircraft. On release, a Pegasus rocket (one of Orbital Science's cruise missile-shaped launchers) accelerated the X-43 to ignition velocity.

A conventional jet engine cannot operate at hypersonic (Mach 5 and beyond) speeds, because its large fan blades and prominent air intake eventually become aerodynamic obstructions. A scramjet has no fan blades—no moving parts at all, in fact. Instead it uses the extreme forward velocity of the host vehicle to force air into a hollow cavity, whose internal shape tapers toward the combustion area, squeezing the air and pressurizing it after it's been

A hypersonic test vehicle is accelerated to ignition speed by a rocket stage. ⋏

Ultraviolet light, in combination with a spray of oil, reveals air flow on the surface of a NASA "blended wing" wind tunnel model. ❯

drawn in. A fine mist of liquid hydrogen fuel is added to the air just before ignition kicks in. A scramjet excels at moving fantastically fast. It is the slow speeds required for take-off and landing that lie beyond its operational range.

In Britain, the Reaction Engines company is developing the Synergetic Air-Breathing Rocket Engine, or SABRE, which essentially changes configuration mid-flight, from jet to rocket engine. On take-off, an internal fan blade sucks up air in the usual manner. Fuel is added from an internal tank. The high-pressure mix is fed to four rocket engines at the rear of the engine assembly, where it is ignited. Then, once the vehicle is pushing into the hypersonic realms, something extraordinary happens. The air intakes are deliberately closed, and a nose cone is pushed forward to streamline the front of the engine housing. At the same time as the external air supply is cut off, an internal oxygen tank is activated. The fan blade compressor inside the engine is bypassed, and the fuel and oxygen mix is fed directly to the rocket engines. SABRE is the brainchild of aerospace engineer Alan Bond, whose principal aim is to create a fully reusable space plane that takes off and lands on a conventional runway, using just one set of engines to accomplish all its goals, whether in the air or in space. Military developers have a particular interest in this kind

of hypersonic vehicle, both for reconnaissance missions and as potential strike weapons. Lockheed Martin's famous SR-71 Blackbird spy plane, developed in the 1960s, was the fastest conventionally powered jet aircraft ever built, routinely achieving speeds of Mach 3 and beyond. A Blackbird could fly from New York to London in one hour and fifty-five minutes. The parallel development of orbiting reconnaissance satellites made the Blackbird redundant by the late 1990s. Nevertheless, Lockheed Martin is promoting a hypersonic unmanned successor, the SR-72. According to Lockheed Martin's program manager Brad Leland, "Hypersonic aircraft, coupled with hypersonic missiles, could penetrate denied airspace and strike at any location across a continent in less than an hour. Speed is the next aviation advancement to counter emerging threats."

Of course, civilian airliner technologists also dream of reducing the duration of long-haul intercontinental journeys. A hypersonic airliner could cut flight times from New York to Sydney, Australia, from twenty-one hours to under three. Many people fail to appreciate that NASA's role is not just the exploration of space. The clue is in the name: the National Aeronautics and Space Administration. It also helps develop new wings and aerodynamic concepts for future air transport systems. Short-term goals include

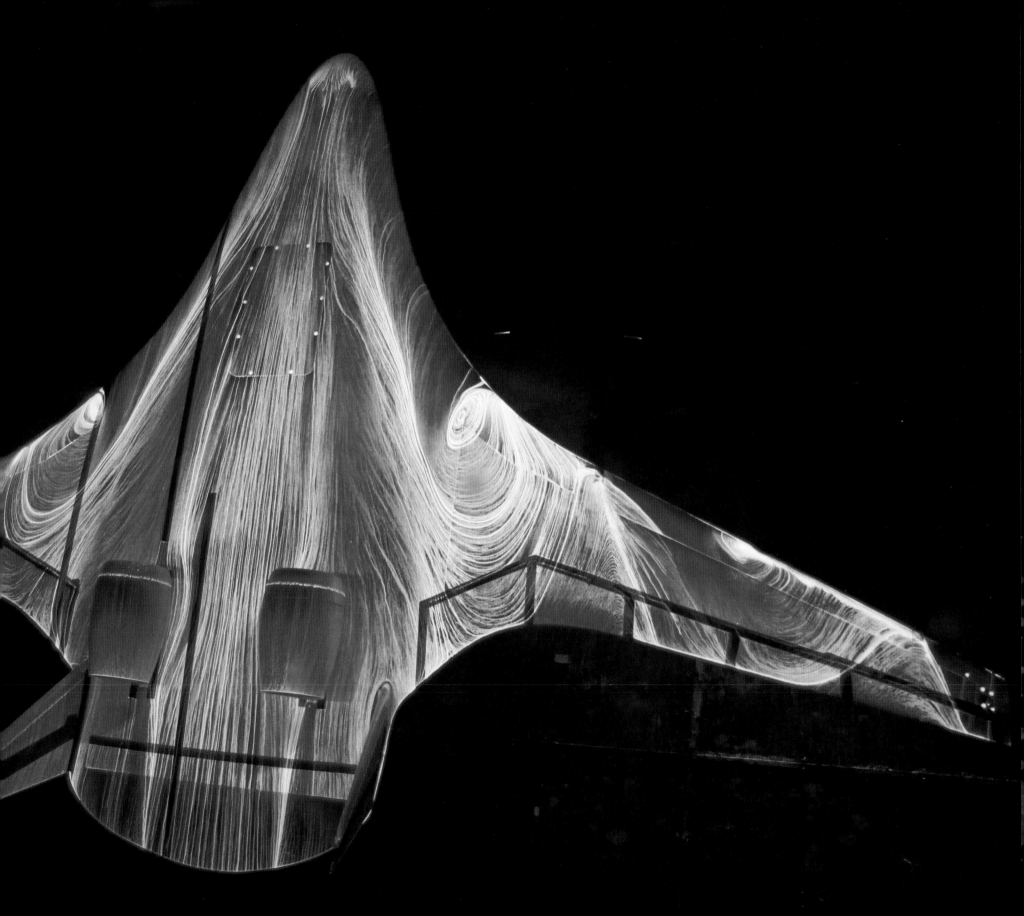

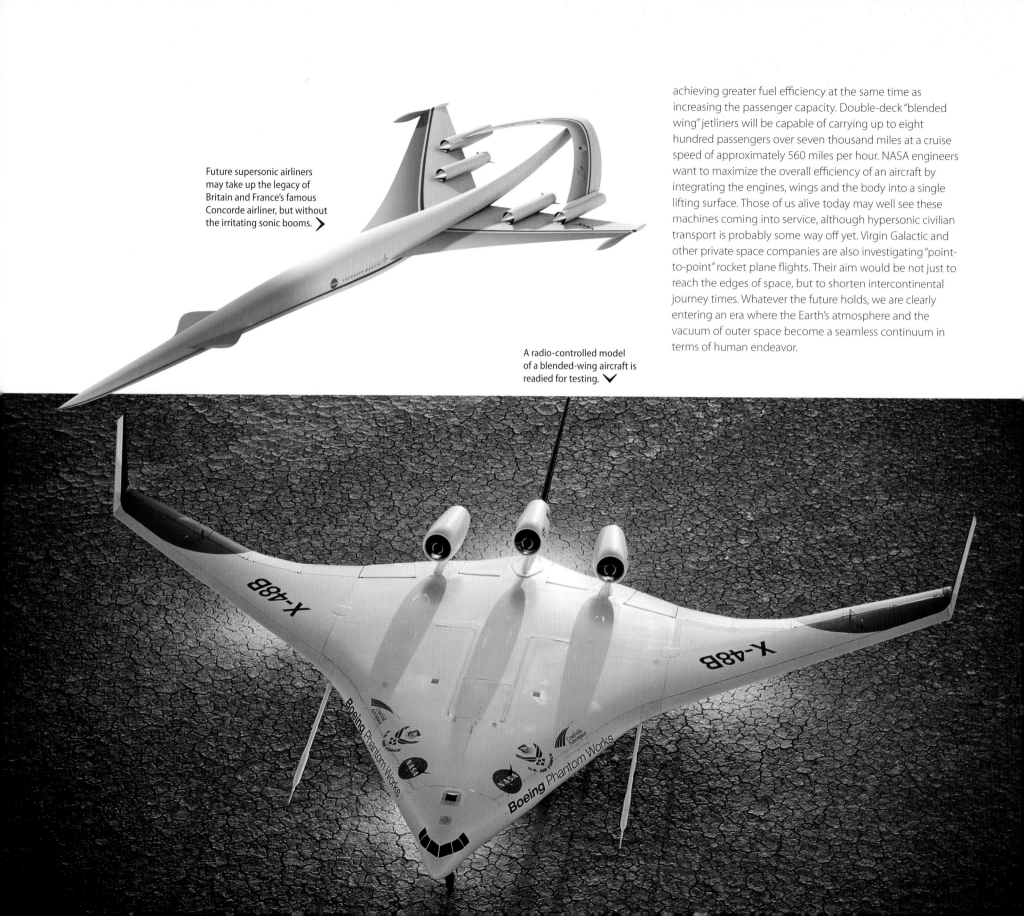

Future supersonic airliners may take up the legacy of Britain and France's famous Concorde airliner, but without the irritating sonic booms. ❯

A radio-controlled model of a blended-wing aircraft is readied for testing. ⌄

achieving greater fuel efficiency at the same time as increasing the passenger capacity. Double-deck "blended wing" jetliners will be capable of carrying up to eight hundred passengers over seven thousand miles at a cruise speed of approximately 560 miles per hour. NASA engineers want to maximize the overall efficiency of an aircraft by integrating the engines, wings and the body into a single lifting surface. Those of us alive today may well see these machines coming into service, although hypersonic civilian transport is probably some way off yet. Virgin Galactic and other private space companies are also investigating "point-to-point" rocket plane flights. Their aim would be not just to reach the edges of space, but to shorten intercontinental journey times. Whatever the future holds, we are clearly entering an era where the Earth's atmosphere and the vacuum of outer space become a seamless continuum in terms of human endeavor.

Felix Baumgartner's capsule ascends under a helium balloon, prior to his record-breaking parachute jump from the uppermost reaches of the atmosphere. ◀

The Slow Route

Hundreds of civilians soon will be able to afford a suborbital rocket flight to the edge of space, and some dozens of especially motivated and well-financed individuals will make it all the way into Earth orbit. Most people never fly above forty-three thousand feet: the upper cruising altitude for jet airliners. However, there is an atmospheric layer, about a third of the way up toward space, that can be reached slowly, gently, and at relatively little cost. World View Enterprises, based in Tucson, Arizona, is developing a near-space travel experience for those who seek greater adventure than merely sitting in an airliner, yet without wanting to reach for space itself. Rocket flight is extremely expensive, and more to the point, it demands a certain level of physical endurance. Sharp accelerations, G-forces, and the risk of motion sickness may not be to everyone's taste. World View is not offering such an adrenaline-fueled ride. This company's craft travels almost painfully slowly, and is powered high into the sky by nothing more ferocious than a giant helium balloon.

A comfortable, fully pressurized flight capsule, accommodating six voyagers and two supervising pilots, is carried aloft by the balloon. Just above the capsule, a semirigid paraglider wing is on standby as a landing system. At the end of each flight, the balloon canopy is detached, allowing the capsule to descend and touch down under reasonably fine aerodynamic control. (The balloon canopy is recovered separately.) The paraglider also serves as a means of emergency escape in case of problems, such as a deflation of the balloon. The capsule's large windows look very much like those of the ISS cupola assembly. They present a superb, panoramic vista. As the balloon climbs to its maximum altitude of one hundred thousand feet, the air thins out dramatically. The characteristic blue haze vanishes, and the curvature of the Earth's horizon is stunningly apparent—all for a fairly modest $75,000 per seat. World View promises that coffee and Internet access will be available during flight. The market for this should be healthy, so long as the system proves safe.

Balloons have carried people into the sky for more than three hundred years. The Montgolfier brothers, Joseph and Jacques, were the first to create a human-carrying hot air balloon. Chemistry teacher Jean-François Pilâtre de Rozier, and an aristocrat, François Laurent d'Arlandes, made the first untethered flight on November 21, 1783. Just two months later, seven passengers were enjoying the view from three thousand feet. Those were a busy few weeks for aviation pioneers. Jacques Charles Robert and Nicolas-Louis Robert made the first untethered ascent with a hydrogen balloon on December 1, 1783. In the twentieth century, ultralight helium has enabled balloons to climb to the uppermost limits of the flyable atmosphere. By the 1950s, a number of scientific and military expeditions to ten thousand feet and beyond, with space-suited crews in pressurized gondolas, were valuable rehearsals for human space flight. In March 1999, the Breitling Orbiter 3 balloon, piloted by Brian Jones and Bertrand Piccard, circumnavigated the globe in twenty days without touching down until the end of the flight.

The logical extreme of a competitive parachutist's ambitions is to fall from as great an altitude as the laws of physics allow—and live to tell the tale. A world record was set in 1960 by US Air Force captain Joe Kittinger, who ascended to an altitude of twenty miles in a gondola attached to a huge helium balloon, then jumped out, briefly reaching a speed of seven hundred miles per hour as he fell. He needed a cooling garment and several insulating layers to protect himself from the heat of reentry: nothing so intense as in the case of a returning space capsule, but uncomfortable nevertheless. As the air thickened at lower altitudes, Kittinger used his body as a crude airfoil, translating some of his downward rush into forward motion

to lose speed before at last opening his 'chute. It took him thirteen minutes to reach the ground. A loose glove put his life in danger just as he was making ready to jump from the gondola, but he was not going to turn back. In the near vacuum of such high altitude, his hand swelled inside the breached glove. The pain was intense. Kittinger gambled on reaching a safe atmospheric pressure fast enough to prevent any permanent depressurization injuries. His gamble paid off, and his hand recovered.

May 2008 was not a happy month for balloonist and parachute adventurer Michel Fournier. At the cost of some $3 million, he took a helium balloon and gondola to a remote prairie wilderness in Saskatchewan, Canada, hoping to surpass Kittinger's record. The sixty-four-year-old French skydiver planned to freefall from an altitude of twenty-five miles. Unfortunately, the balloon ripped free of its cables and soared off during the inflation process, leaving Fournier and his gondola sitting on the ground.

The technology for helium balloon ascents is already well known. The challenge will be to ensure that a capsule can be recovered if the main balloon fails. ❮

World View's pressurized gondola surely will deliver a fabulous view for its passengers. ⌄

> "WITH NO SPECIAL TRAINING, YOU'LL EXPERIENCE A THRILL LIKE NO OTHER: ONE THAT FOR MORE THAN HALF A CENTURY HAS BEEN RESERVED SOLELY FOR ASTRONAUTS"
>
> World View Enterprises, Inc., 2014

World View's gondola, suspended in a high-altitude region of the atmosphere, will deliver an experience almost like being in space, except that its passengers will not be weightless. ❯

THE MAN WHO FELL TO EARTH

Joe Kittinger's record was broken by Austrian skydiver Felix Baumgartner on October 14, 2012. His capsule and balloon, sponsored by the Red Bull GmbH (maker of the world's most popular energy drink) ascended twenty-four miles. After a final check of his equipment, Baumgartner depressurized the capsule, opened the hatch, and jumped, wearing a space suit whose claustrophobic tightness had nearly caused him to resign from the project many months earlier. By the time he had fallen to ninety-eight thousand feet he was moving faster than sound. After four minutes of free fall, he had descended to eight thousand feet, where he could deploy his 'chute. Nine minutes after jumping out of the capsule, he was standing on a patch of scrubland just outside Roswell, New Mexico, and happily opening his visor.

Baumgartner's mission "to the edge of space" did not make him an astronaut. His gondola would have had to ascend three times higher for that label to be appropriate, but who's counting? The human experience of such great heights is sufficiently space-like not to make much difference. So-called "space diving" is likely to grow in popularity.

Baumgartner is loaded into his capsule, wearing a space suit to protect him from the near-vacuum conditions at the top of his ascent. ∧

The Red Bull drinks company was a major sponsor for Baumgartner's jump. We can expect further commercial adventures in, or near, space. ＞

Baumgartner pauses for a moment on the edge of a precipice... Would you be brave enough to make that jump? ❯

Felix is safely on the ground after a nine-minute fall. ❮

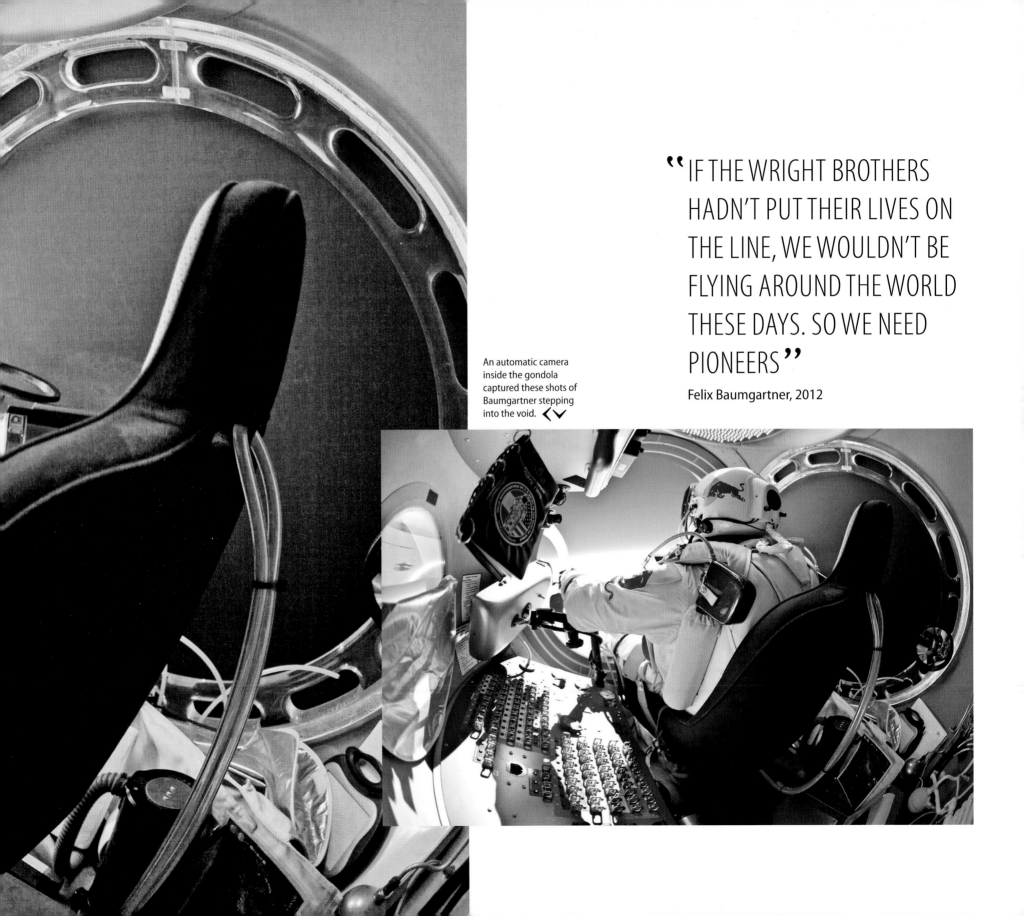

> "IF THE WRIGHT BROTHERS HADN'T PUT THEIR LIVES ON THE LINE, WE WOULDN'T BE FLYING AROUND THE WORLD THESE DAYS. SO WE NEED PIONEERS"
>
> Felix Baumgartner, 2012

An automatic camera inside the gondola captured these shots of Baumgartner stepping into the void. ◀∨

03.

The 2013 movie *Elysium* suggested a social division between rich space colonies and poor communities on Earth. ∧

A gigantic O'Neill Cylinder could provide comfortable conditions for thousands of people. ≪

Look up at the right time and place, and you will catch a glimpse of the International Space Station, by far the largest space structure that we have ever sent aloft, arcing across the sky. The multimodule ISS encloses a greater pressurized volume than a 380-seat Airbus A340 and the area of its solar arrays and structural trusses would cover an American football field (including the end zones). Given that ISS has been operational for nearly two decades, it may seem strange to point out that, as of today, no one really knows what space stations are for.

The basic concept owes its origins to the great Russian pioneer of astronautics, Konstantin Tsiolkovsky (1857–1935) who worked on detailed ideas about rockets at the very dawn of the twentieth century. As early as 1903

he proposed a huge habitable cylinder, spinning on its axis and containing a greenhouse with a self-supporting ecological system. His ideas were informed by cosmism, a mystical fusion of religious and technological thinking that appealed to many Russian thinkers at that time. In essence, our ascent to realms beyond the Earth was supposed to elevate the human condition. Space habitats were just one technology that would assist our evolutionary progress.

The first detailed engineering proposal for a space station appears to have been drawn up by Hermann Noordung, a military engineer from the days of the Austro-Hungarian Empire. His short book *The Problem of Space Travel* (1927) is still an influential touchstone for aerospace engineers. We know very little about Noordung's life, except

Every day for the last two decades, humans have been living in space. The challenge, now, is to increase that off-world population and enlarge their habitats.

ISLANDS
IN THE SKY

that he died tragically young of tuberculosis just two years after publishing his book. Fortunately his space station proposal survives in all its prescient detail. He conceived a "habitation wheel" whose gentle rotation provides its crew with artificial gravity. Airlocks and safety bulkheads are all described, along with a huge parabolic dish that collects and focuses sunlight for power. The docking airlock rotates at the same rate as the station, but in the opposite direction, maintaining its position relative to the Earth's horizon. This allows rocket ships to approach without themselves having to tumble. Two years after Noordung, the British molecular biologist J. D. Bernal wrote *The World, the Flesh and the Devil* (1929), in which he proposed self-supporting space "Worldships," each capable of housing

many thousands of inhabitants. In the early 1940s, the British futurologist and science fiction writer Arthur C. Clarke helped develop radio navigation systems for Allied aircraft returning from raids over Germany. At the end of World War II he recognized that his cherished dreams of space flight would never be achieved unless governments and industrial investors could be persuaded to see the economic benefits of large-scale rocket programs. In October 1945 he published an article in a British electronics magazine, *Wireless World*, essentially outlining the modern concept of geosynchronous communications satellites.

The only difference between Clarke's ideas and the global satellite network that we see today is that he was thinking in terms of bulky and unreliable 1940s

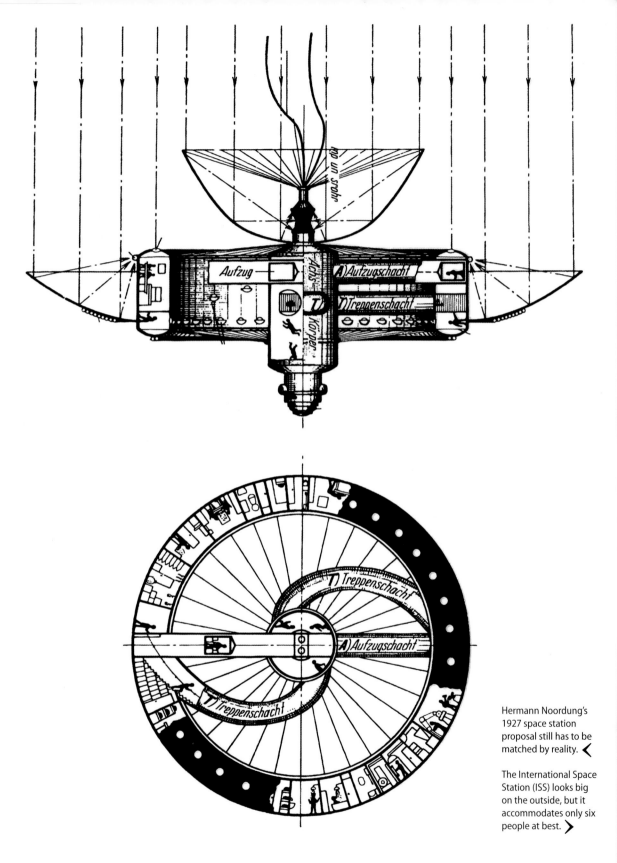

Hermann Noordung's 1927 space station proposal still has to be matched by reality. ◄

The International Space Station (ISS) looks big on the outside, but it accommodates only six people at best. ►

communications equipment assembled from glowing thermionic valves, circuit breakers, and wiring boards. With this fact in mind, he suggested that his "Extraterrestrial Relays" would have to be staffed by human engineers, recruited in all probability from a significant pioneer of technology in those days: the General Post Office, a government-backed organization which, for several decades, was responsible for all mail, telegraph, and telephone services in Great Britain. Clarke's orbiting outpost had a radio room, a small library, a kitchen, a surgery, cabins for the crew, and of course, separate and more luxurious quarters for the station's commander. (This was a very *English* space station.) The swift development of microchip technology soon eliminated any need for Post Office engineers in space.

One possible space station use that has never quite gone out of fashion is that of a solar power harvesting platform. The collected energy could be converted into a microwave beam fired down to Earth. On the ground, microwave dishes could capture the beam and convert it into usable electricity. The advantage of this technique is that the Sun's energy is free and limitless. The technology for a space power station has existed for many decades. The disadvantages, however, are daunting. To make any significant contributions to Earth's electrical appetites, a space power system would require a huge collector array in orbit and several square miles of receiver panels on the ground. This could create environmental problems.

Another possibility is to concentrate the beam onto just one small ground receiver. But that strategy presents a different problem: what happens if the beam strays from its target? Could such a machine be misused? It is a real pity that we cannot yet make better use of the Sun to produce electricity on the ground, but so far, unfortunately, no one has come up with an appropriate balance of benefits versus costs. For now, the prospect of spending tens of billions of dollars on more efficient terrestrial energy sources seem to outweigh the hundreds of billions it might take to build a space power station.

Where once the gentle spinning of space stations was seen as a way of making life more or less bearable for their inhabitants, near-weightlessness is now regarded as the main justification for building them in the first place.

This is why the first Russian and American space stations, and of course, ISS, look nothing like the visions of the early space theorists. Space stations can be as haphazard and inelegant as they like, because they do not have to spin. Unfortunately, a zero-gravity (or, more accurately, microgravity) environment is not so great for the human body. Astronauts experience calcium loss and skeletal weakening quite similar to osteoporosis, a common problem for aging people. In space, symptoms accumulate over days and weeks, even in the youngest and fittest astronauts, while on Earth it usually takes many years for osteoporosis to become serious. Astronauts protect themselves with vigorous exercise, and their bones recover fully once they are back home.

ISS originated from what was to have been a purely American enterprise. In January 1984, President Ronald Reagan invited like-minded European partners to join the project, known at the time as Space Station Freedom. The dissolution of the Soviet Union in the 1990s resulted in a substantial redesign of the station and a new emphasis on its political justifications. Reagan's somewhat provocative name was dropped, for there was no longer any prospect that the Soviet Union might remain a rival in orbit. The Soviet Union no longer existed. Russian and American space crews began to collaborate aboard the Mir (Russia's last independent space platform), and the proposed new station was redesigned to include Russian modules. It was renamed Alpha before finally becoming ISS. In fact, the first functioning module, Zarya ("Dawn") was placed into orbit by a Russian Proton rocket in November 1998.

At their height, the Soviet space and missile programs had employed upwards of half a million people. As the old empire collapsed, Western leaders saw the benefits of keeping talented Russian rocket engineers busy, just in case their expertise leaked out to unsuitable countries. Of all the purposes that a space station might have achieved, what could have been better, or more unexpected, than cementing international relations among former enemies? ISS evolved into a unique experiment, not just because of its many science payloads, but also in terms of human and societal relationships. This centerpiece of our generation's human space flight adventure should remain operational until at least the year 2024.

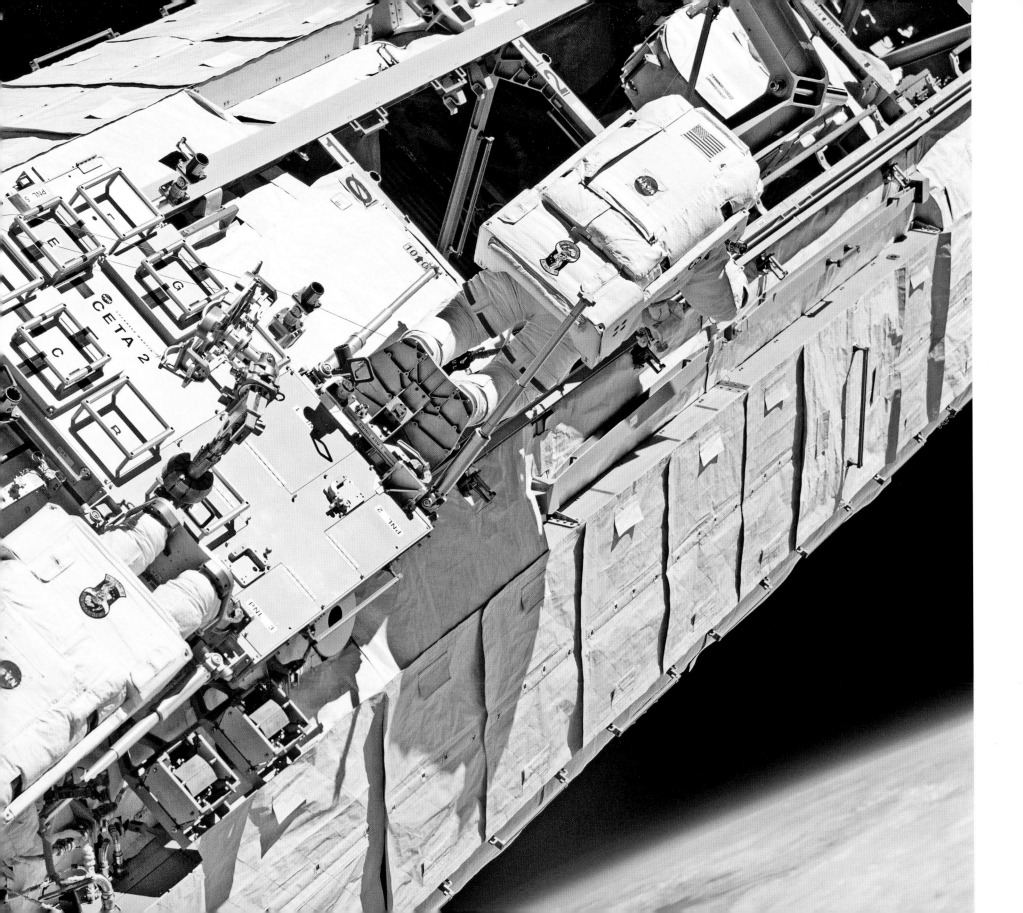

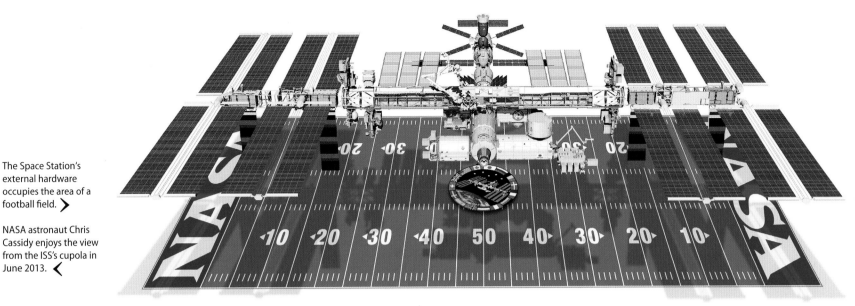

The Space Station's external hardware occupies the area of a football field. ❯

NASA astronaut Chris Cassidy enjoys the view from the ISS's cupola in June 2013. ◀

SCIENCE ON ISS

Material & Fluid Science
Whenever a substance is melted and cooled here on Earth, gravity has a strong influence over what happens. The cooler, denser regions and the heavier chemical elements of molten liquid mixtures tend to sink to the bottom. Experiments in microgravity point the way to eliminating these problems and creating materials of exceptional uniformity throughout their structure.

Fundamental Physics
When solids turn to liquid, and liquid turns to gas, these three familiar states are distinct from each other, but the transition from one state to another creates a fleeting, little-understood state of matter known as "phase transition." In space, this delicate area of atomic physics can be studied in an ideal environment where no forces other than the phase transition itself are at work. ISS is also studying cosmic rays: energetic subatomic particles arriving in a constant barrage from the deepest realms of space.

Biology
How do microorganisms and insects sense gravity and respond to it? Taking away the effects of gravity helps us work out how plants and other organisms distinguish between "up" and "down," and how roots and shoots grow in the appropriate directions. Miniature centrifuges aboard ISS create variable artificial gravity fields, testing the threshold sensitivities of young plants.

Biomedicine
This is the area of research that might just make us grateful for ISS one day. In space, bones shed calcium at an alarming rate. Astronauts make great orbiting specimens because they temporarily suffer a condition similar to osteoporosis, which is usually associated with aging. Nearly 30 million Americans suffer from osteoporosis. The total costs of treatment amount to $20 billion a year, or one fifth of the entire ISS construction budget.

As a tool for reinforcing geopolitical alliances, ISS probably has been worth every cent. When it comes to tangible scientific rewards, the jury is still out. One of the problems with onboard experiments is that they cannot easily be replicated and verified by scientists on Earth, let alone applied to new forms of manufacturing or research down on the ground. As a consequence, NASA has tended to shift ISS's mission away from general science and toward a more specific area: finding out how humans survive and cope while in orbit for long periods. In essence, ISS is a rehearsal for future deep space missions. If we can figure out how to keep people healthy for a year or more at a stretch aboard ISS, this will go a long way toward preparing for missions to Mars.

Critics of ISS say that humanity has already accumulated half a century's worth of human space flight data. Some are anxious to get on with exploring deep space, acutely aware of the fact that the public senses no great glory in occupying low Earth orbit for year after year. Furthermore, professional astronaut missions are inherently risky. When these dangers are attached to a goal that is seen as worthwhile, such as landing on the Moon or repairing the Hubble Space Telescope, there is broad public acceptance that the benefits are worth a degree of risk.

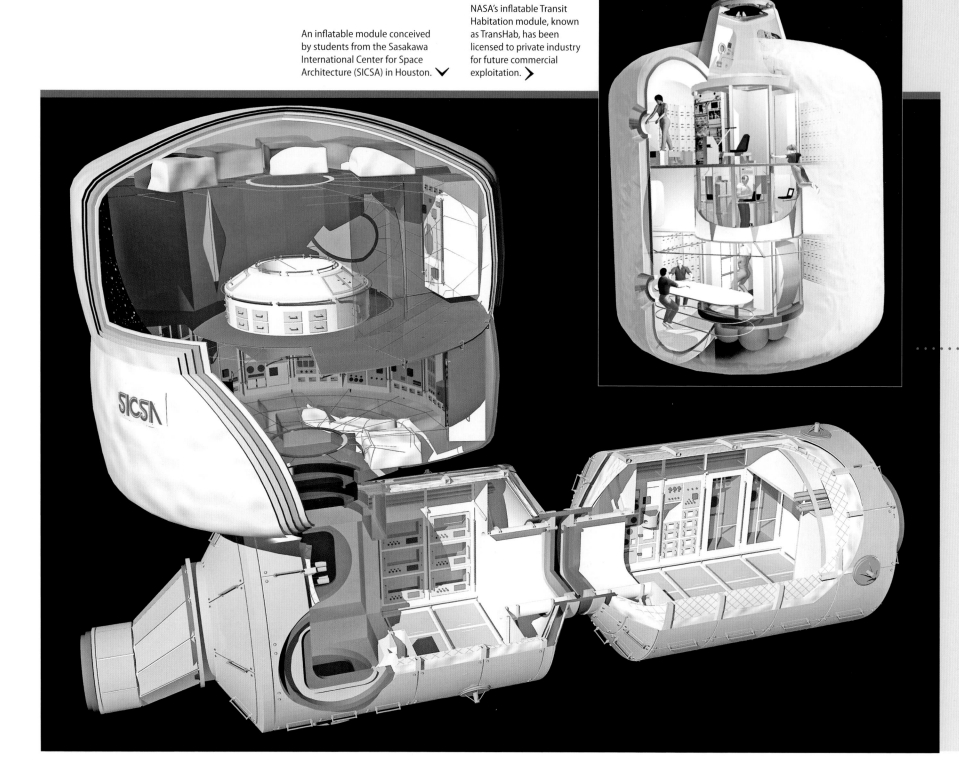

An inflatable module conceived by students from the Sasakawa International Center for Space Architecture (SICSA) in Houston. ∨

NASA's inflatable Transit Habitation module, known as TransHab, has been licensed to private industry for future commercial exploitation. ❯

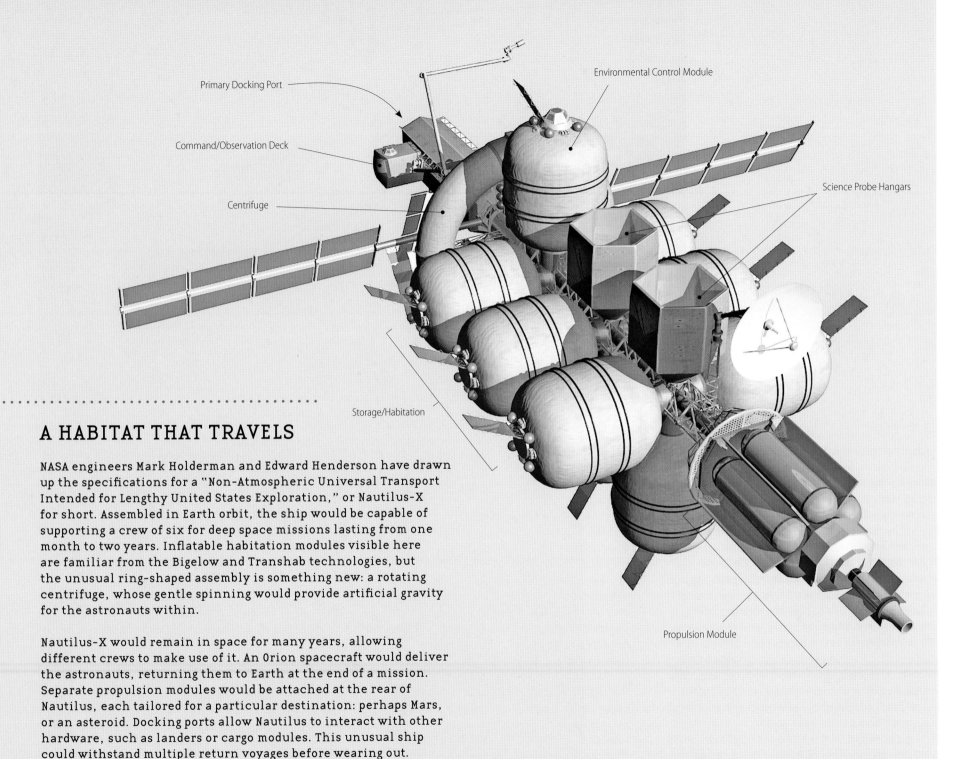

Primary Docking Port

Command/Observation Deck

Centrifuge

Environmental Control Module

Science Probe Hangars

Storage/Habitation

Propulsion Module

A HABITAT THAT TRAVELS

NASA engineers Mark Holderman and Edward Henderson have drawn up the specifications for a "Non-Atmospheric Universal Transport Intended for Lengthy United States Exploration," or Nautilus-X for short. Assembled in Earth orbit, the ship would be capable of supporting a crew of six for deep space missions lasting from one month to two years. Inflatable habitation modules visible here are familiar from the Bigelow and Transhab technologies, but the unusual ring-shaped assembly is something new: a rotating centrifuge, whose gentle spinning would provide artificial gravity for the astronauts within.

Nautilus-X would remain in space for many years, allowing different crews to make use of it. An Orion spacecraft would deliver the astronauts, returning them to Earth at the end of a mission. Separate propulsion modules would be attached at the rear of Nautilus, each tailored for a particular destination: perhaps Mars, or an asteroid. Docking ports allow Nautilus to interact with other hardware, such as landers or cargo modules. This unusual ship could withstand multiple return voyages before wearing out.

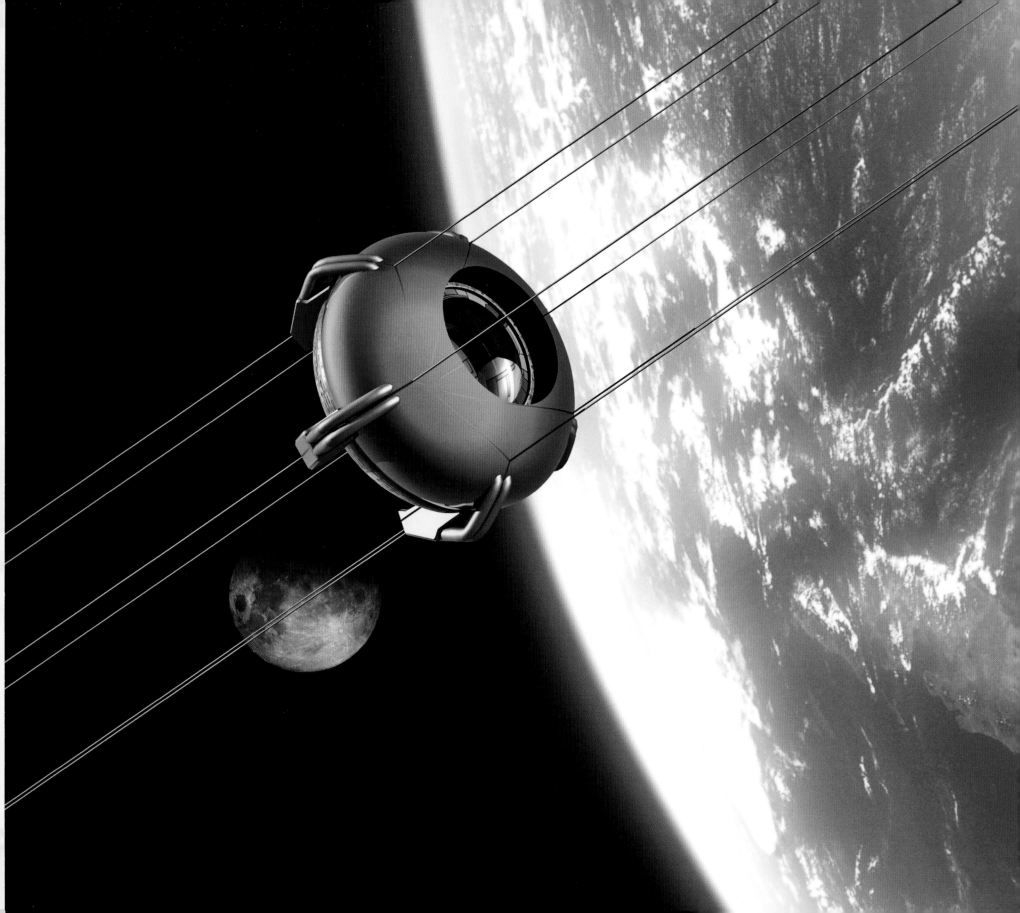

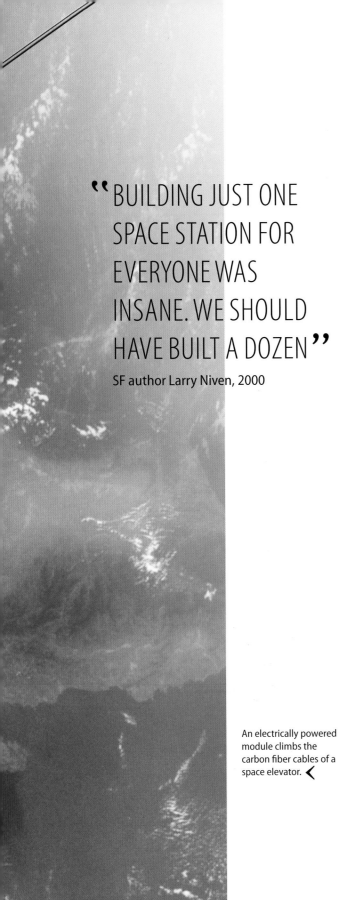

> ## "BUILDING JUST ONE SPACE STATION FOR EVERYONE WAS INSANE. WE SHOULD HAVE BUILT A DOZEN"

SF author Larry Niven, 2000

An electrically powered module climbs the carbon fiber cables of a space elevator. ◀

NASA to award Bigelow Aerospace a $17 million contract, with the aim of testing a small prototype, the Bigelow Expandable Activity Module (BEAM) docked to ISS.

Space Elevators

Those familiar with Arthur C. Clarke's science fiction novels will surely think of him as the champion of the space elevator. In his 1979 novel *The Fountains of Paradise*, a mountaintop in a fictionalized version of his beloved Sri Lanka is linked to a platform in geosynchronous Earth orbit via a thin cable, twenty-two thousand miles long, made from "continuous pseudo-one-dimensional diamond crystals." Electrically driven pods climb this eerie monorail, reaching space with none of the wasteful sound and fury associated with today's rocket technologies. Clarke kept his fingers crossed that new developments in molecular carbon technology would help turn his fictional notion into fact. In 2003 the University of Kentucky created something close to what he had imagined: an immensely strong three-mile length of carbon cable, thinner than a human hair, woven from microscopic nanotubes. Carbon nanotechnology is a booming sector today.

Clarke never claimed to have invented the basic principle of the space elevator. As early as 1895, the prescient Russian rocket theorist Konstantin Tsiolkovksy described an artificial "beanstalk" reaching up to "a celestial castle" at just the same geosynchronous orbital altitude later suggested by Clarke. In 1960 another Russian, Yuri Artsutanov, described how to drop a cable down to Earth from a geosynchronous platform. He also added that the drag of the tether's weight would have to be centrifugally counterbalanced by an equivalent length of cable stretching beyond the platform and into space.

By the mid-1960s NASA was experimenting with some of these ideas in earnest, albeit on a tiny scale. The space agency's goal was to create artificial gravity by spinning two spaceships, linked by a tether, around a common center of gravity. Astronauts on the last two Gemini missions tethered their capsules to docked Agena target vehicles. Real life proved to be messier than the mathematical models. Unpredictable forces twisted the tethers into strange loops, while the Gemini and Agena vehicles tugged and fought each other in a vexing dance. In September 1966 Gemini 11 astronauts Pete Conrad and Dick Gordon achieved a rotation rate of one revolution

every six minutes, creating a detectable artificial gravity and satisfying the honor of all concerned.

Electrically conductive tethers have even been deployed from simple sounding rockets to probe the charged environment of the ionosphere (the upper region of atmosphere where the auroral lights flash and flicker), while various agencies concerned by space debris are studying how to remove unwanted satellites from orbit by commanding them to unspool thin cables. Over a period of weeks or months, electromagnetic drag effects (the Lorentz force) cause their orbits to decay. We are a few decades—and perhaps a century—away from a Clarkean space elevator. Even so, the concept is stimulating a great deal of research already.

Cities in Space

Today it is hard to imagine a time when U.S. Senators listened in rapt attention while a charismatic lecturer argued for the construction of giant orbiting habitats as a way of easing environmental pressures on Earth. The structures, at least two miles long, would support thousands of people, all living in leafy suburbs. In January, 1976 it was possible for Gerard K. O'Neill, a physics professor at Princeton University, to talk about space colonies without sounding like a dreamer. Whatever happened to an idea that once captivated the world?

An accomplished scientist, O'Neill was already renowned for creating a machine that temporarily stores high-energy particle beams in magnetic fields so that they can be released, on command, to smash into other beams: a crucial technology behind particle colliders. However, by 1969 he was experiencing the classic inventor's disenchantment with senior management at the institutions where he worked and was on the lookout for fresh inspiration. He found it in the wake of *Apollo 11*'s lunar touchdown. As he recalled for a NASA interview a few years later, "It just seemed to me that to be alive at that time, and not to try to take part in that unique event in human history, the first breakout from the planetary surface, would be something I would regret forever."

At first, O'Neill incorporated space exploration just as a theoretical concept to stretch the imaginations of his students. "This was the peak time of disenchantment with anything in science and engineering. The students who were good at it felt very defensive, because all of thruster fuel. Inside the cylinders, three lengthwise land areas alternate with three window strips, illuminated with mirrors that open and close to create a familiar cycle of day and night. Each cylinder rotates about forty times an hour, simulating Earth's gravity. A total land area of five hundred square miles accommodates several million people. An outer ring of greenhouse pods provides plenty of food.

O'Neill's research assistant, Eric Drexler, proposed simple inflatable structures that would be rotated in front of a metal vapor spray, similar to the devices used to dope microchips with their many different layers of conductor. Wires or rods of raw materials would be fed into the system, melted into vapor, then deposited at relatively low temperatures on the surface of the plastic template. He knew what he was talking about. Today Drexler is one of the world's leading pioneers of nanotechnology, the technique that allows objects of any size to be engineered upward from the molecular level.

Within each gigantic shell, O'Neill wanted to build a leafy paradise. "I had no desire to just invent a space station. It had to be beneficial for a lot of people. It had to look an awful lot like the Earth." But how was all this to be funded? The colony's initial purpose was to house the builders of the ultimate nonpolluting power system for Earth itself. According to O'Neill, solar energy was the obvious way to go. Gigantic orbiting arrays convert sunlight into a microwave beam, while, on Earth, fields of collector dishes in unpopulated deserts gather the microwave energy and reconvert it into electricity. This would pay for a colony in less than twenty-five years, he calculated.

O'Neill never once considered the space community's usual goal of colonizing other worlds. "We side-tracked that very quickly because Mars, or any other alternative planetary surfaces, are fairly unpleasant options. They are the wrong distance from the sun, and have the wrong gravities." He saw no sense in giving up the solar energy available in near-Earth space in favor of the cold, weak sunlight available on a distant planet.

On January 19, 1976, O'Neill testified before the U.S. Senate Subcommittee on Aerospace Technology and National Needs. The hearing room was packed to overflowing as he held out the prospect of an "inexhaustible energy source" for an initial outlay of no more than three times the cost of Project Apollo. In its subsequent report, the committee said that space habitats were "potentially feasible" and deserved more study. Above all, "methods for space-based generation of electricity, using energy from the sun, should be developed as a significant contribution to the fossil fuel dilemma."

NASA was not in a position to build any hardware, but made some funding available for theoretical studies, including conferences at Stanford University during 1975 and 1976, where O'Neill's ideas gained real academic stature. The only serious flaw in the plan—and perhaps one that NASA was happy to fudge—was O'Neill's casual approach to the first crucial crews and payloads that would need to escape from Earth in order to get the manufacturing processes under way. Everyone expected too much from the Space Shuttle, which had yet to make its first flight. The lunar mass driver, O'Neill reckoned, would call for "a year's worth of shuttle flights." No one among his team gave serious thought to the problem of launch vehicles, let alone the differences between an Earth-orbiting shuttle and the hardware required for the first post-Apollo touchdowns on the moon or the initial cramped quarters for the first colony engineers. "We would use the shuttle's external tanks to make modular living-quarters for use in low and high orbit and on the lunar surface," O'Neill wrote, somewhat optimistically.

Just a few minutes' drive from O'Neill's offices at Princeton, a firm of consultants was studying the shuttle's economic benefits for NASA. They suggested that thirty-nine flights per year, with up to five hundred flights conducted between 1978 and 1990, transporting humans and cargo alike in the same vehicles, would make the shuttle a winner. O'Neill took this on faith. Of the many wonderful visualizations in *The High Frontier* made by artist Don Davis, none specify customized launch hardware. With Apollo so recently triumphant, it seemed obvious to most people that NASA would come up with some solution.

O'Neill died of leukemia on April 27, 1992, at the age of sixty-six. Thousands of space enthusiasts lost an

A fusion of mechanical and biological life support systems may lead to "living spaceships." ➤

BUILDING PARADISE IN SPACE

Project Persephone is a collaborative research effort led by scientists in America, Italy, and the Netherlands. Their aim is to investigate how future biological technologies might enable a space habitat to become truly self-sustaining and even, in some senses, "alive" in its own right. An internal ecosystem would generate air, water, and food for human occupants. Named in honor of the ancient Greek goddess of vegetation, this project also tells us a great deal about the fragility and interdependence of ecosystems on Earth.

inspirational figure that day. But his ideas haven't gone away. It has become possible once more to talk about harnessing solar energy in space, albeit on a smaller scale than O'Neill would have liked. Perhaps the private sector will take the lead. Entrepreneur and space activist Bob Werb collaborated with O'Neill in the late 1970s, before founding another high-profile lobbying group, Space Frontiers, in the 1980s. "The value of Gerard's vision is, if anything, more certain today than ever. The proposed new direction for NASA is fundamentally O'Neillian, because it looks toward enabling technologies that will benefit all of us, instead of just small groups of astronauts." With private space station pioneers such as Bigelow in mind, Werb suggests that "the scale of Island Three is still way beyond our current capabilities, but small-scale settlements are going to happen very soon."

Artwork from the 1970s depicts the interior of an O'Neill Cylinder space colony. ⋀

An alternative proposal, known as the Stanford Torus, is shaped like the inner tube of a car tyre. ❯

04.

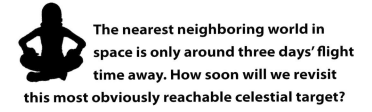

The nearest neighboring world in space is only around three days' flight time away. How soon will we revisit this most obviously reachable celestial target?

DESTINATION
MOON

Moonbase Alpha, in the 1970s TV series *Space: 1999* (above) was inspired by *2001: A Space Odyssey* (inset). We are waiting for reality to catch up. ◄

Space industry insiders sometimes use a sarcastic acronym, ABV, which stands for Air-Brushed Vehicle: something that looks convincing when sketched by an engineer (whether in paper or in pixels) but never gets built. Many ABV examples have been portrayed landing alongside equally convincing Moonbases. In January 2004, President Bush announced that astronauts should return to the Moon in the coming decade and establish a permanent base as practice for future missions to Mars. This political momentum galvanized space designers within NASA, ESA, and the Japanese Aerospace Exploration Agency (JAXA). Some of those ABVs looked as if they might fly off the page and reach the Moon. Then President Barack Obama's administration put the lunar exploration aspects on hold while signaling that a mission to rendezvous with—and perhaps even capture —a small near-Earth asteroid might be a more achievable goal than attempting to land on Mars. As has happened so often in the past, NASA's fortunes

"GOING BACK TO THE MOON, YOU ACCELERATE YOUR ABILITY TO GO ANYWHERE ELSE. AND IT'S ONLY THREE DAYS AWAY"

Astronaut Harrison Schmitt, 2012

New space suits and roving vehicles are being tested on Earth. ◀

The evolution of rover technology, from Apollo's open two-seater to a multi-purpose adaptable chassis supporting a pressurized crew compartment. ▶

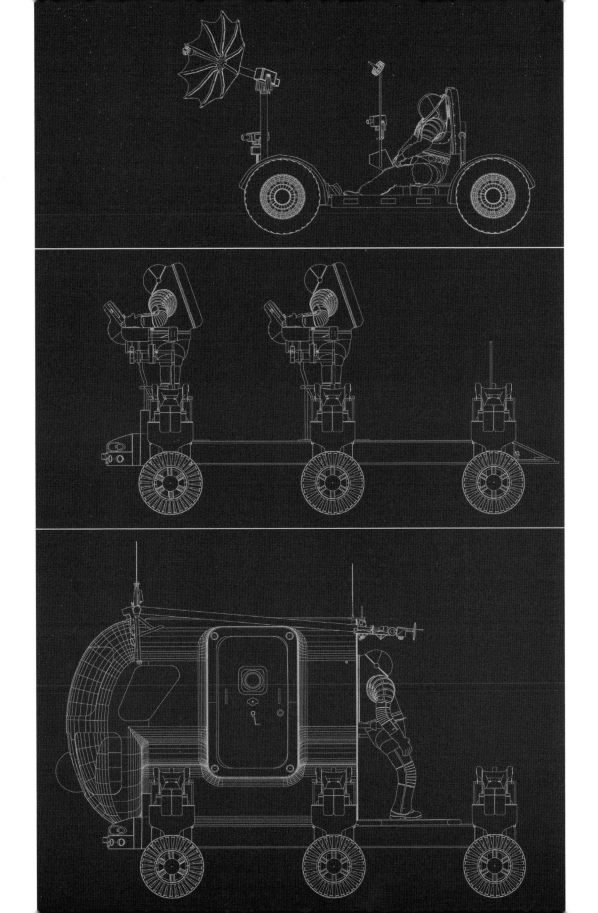

drifted on the ebb and swell of political tides and national economic fortunes.

Whatever the eventual schedule for a human return to the Moon turns out to be, one of the most obvious problems already may have been solved: where to place a Moonbase. In April 2005, a team at Johns Hopkins University Applied Physics Laboratory, led by British-born Ben Bussey, released a paper in the international science journal *Nature*. Data gathered in 1994 by the unmanned Clementine space probe was analyzed to produce a map of the Moon's north pole, revealing the percentage of time that the surface is illuminated by the Sun during the lunar day. This may sound like just a technical scientific inquiry, but the results yielded possible prime locations for a future lunar habitat.

In full daylight the lunar surface can reach blistering temperatures of 212 degrees Fahrenheit (100 degrees Celsius), while at night it plunges to -292 degrees F (-180 degrees C). This environment would place great stresses on a potential base. The long nights (two terrestrial weeks) would also play havoc with solar power supplies, but Bussey's team found at least one place that stays almost permanently in daylight: a spot where human-built structures could be maintained all year round at a relatively benign temperature of -58 degrees F (-50 degrees C). Polar science bases on Earth cope with harsher temperatures than that. The magic spot is on the rim of Peary crater, close to the Moon's north pole, named after the U.S. explorer Robert Peary, who in 1909, after a trek of thirty-seven days, reached Earth's North Pole. (Although some critics suggested he may have ended his journey at the wrong place.)

Another line of inquiry was directed at an opposing phenomenon: lunar regions shrouded in permanent shadow. In 1998 a neutron spectrometer aboard NASA's *Lunar Prospector* spacecraft scanned the topsoil for hydrogen-rich compounds. High readings in the flanks of some deep craters strongly suggest the presence of water ice, shielded from evaporation because the Sun's warmth never hits those sites. NASA believes that some of the ice could be mined for drinking water, and the rest broken

A prototype Multi-Mission rover is put through its paces in support of a desert geology team. ◀

down, by means of solar-powered electrolysis, into oxygen for life support and hydrogen for fuel. More cautious voices have warned of the immense difficulty of mining hardened permafrost in pitch darkness at the bottom of steep craters, but we will never know until we try.

Of course, lunar ice is just a potential utility resource. It is not, on its own, a reason for going back to the Moon. What about extracting rare metals or digging for valuable minerals? The dusty topsoil (regolith) contains plenty of oxygen bound up in its minerals. There are also high concentrations of silicon, aluminum, iron, magnesium, and titanium, as well as smaller amounts of chromium, manganese, and sulfur. These elements can be extracted and used as construction materials. Supporters of lunar exploitation insist that the Moon's natural resources can be profitably exploited for terrestrial purposes, too. Opponents say that there are relatively few native materials of sufficient abundance, accessibility, and value to merit their extraction and export. The costs would be prohibitive. Until someone goes back to the Moon and make an initial prospecting survey, one cannot be sure.

Some studies do suggest that lunar helium-3, deposited in the topsoil by charged particles in the solar wind, could be worth the effort of refining it. Just twenty tons of it could power the entire United States for a year, as long as nascent fusion reactor technology can be made to work. According to some estimates, the Moon possesses millions of tons of helium-3—enough to power the entire Earth for thousands of years. Extracting this lightweight and fragile treasure from the topsoil would be complex and expensive, but every shipload sent back to Earth could be worth several millions of dollars per pound. Naysayers worry that each ounce of this treasure could be collected only by sifting through thousands of tons of lunar soil. No one will know for certain until further investigations are conducted.

A lunar base may be better for pure scientific research rather than commerce. The Apollo missions investigated

A 3D printing system lays the foundations for a dome habitat. ⋀

Renowned architect Sir Norman Foster suggests that lunar bases can be "printed" from jets of processed topsoil. ⟨

PRINTING A MOON BASE

Building the infrastructure for a lunar base need not be complicated. Living quarters can be constructed by linking together Bigelow Aerospace's inflatable modules and covering their roofs in protective blankets of lunar topsoil. This is a simple method of shielding the interior against micrometeorites, while keeping control over temperatures. New developments in 3D print technology are intriguing. British architect Norman Foster is just one among several practitioners investigating how to build a base's outer layer by gathering loose topsoil, grinding it to a fine powder, depositing it from a rotating nozzle, and heat-sealing it into place, creating a dome-shaped shell.

Long-range wheeled rovers will be essential. Docking hatches will allow multiple rovers to link up and create temporary exploration camps far afield. The rovers will be fully inhabitable extensions of the main base. They will have exterior robotic grappling tools, enabling scientists to study rocks and soil samples without always having to climb into their space suits.

just a tiny fraction of the Moon's surface, but for reasons of safety and fuel economy, pioneering astronaut-explorers could not access the lunar mountains, nor descend into ravines or deep craters. Landing sites were chosen from the relatively flat "seas" of the equatorial belt. Indeed, there is a vast amount of terrain still to explore, including the polar regions and the whole pockmarked far side of the Moon, where no human—or machine—has yet safely touched down. Since the dawn of the space age half a century ago, many astronomers have dreamed of building telescopes and radio dishes inside deep craters where they are shielded from solar light pollution. The lunar far side could also screen sensitive radio instruments against Earth's babble of radio noise.

It may be that we'll just go back to the Moon for the sheer sake of simply doing so. Just two day's flight time away, it is a tempting destination for robotic and human visitors alike. The descent stages of lunar modules from six Apollo missions are perfectly preserved at their landing sites. The astronauts' footprints in the surrounding soil look as fresh as they did when they were imprinted four decades ago. There is no wind on the airless Moon to disturb them and no rain to wash them away. Small robot rovers are being planned by private companies who want to photograph the *Apollo 11* site up close, but without touching anything. It will be headline news when one of them succeeds, but it will also useful for future Moon base designers to see just how well (or badly) those Apollo relics have lasted in the harsh lunar environment.

What to Wear

A space suit is essentially a miniature spacecraft, complete with everything that an astronaut needs for survival: oxygen to breathe, water to drink, and a system for controlling the interior temperature. A backpack holds the oxygen and water, along with a radio, a battery, and a water-cooling pump. Astronauts wear an inner garment, similar to long johns, threaded with thin hollow tubing. Water from the backpack cooling system circulates through the tubes, absorbing excess body heat. Then the water is pumped around again so that the heat can be lost into space from a small radiator on the backpack.

The first space suits, developed in the late 1950s, were designed mainly to protect astronauts in case their capsule sprang a leak. On March 18, 1965, Soviet cosmonaut Alexei Leonov became the first man to "walk" in space. He squeezed into the cramped flexible airlock of his Voskhod capsule and pushed himself outside. For ten minutes he enjoyed the exhilarating sensation of space walking, and then began to pull himself back into the ship, only to discover that his suit, at full pressure, had ballooned outward in the vacuum of space so that he could no longer fit into the airlock. Dangerously exhausted by his efforts, Leonov had to let some of the air out of his suit to collapse it so that he could squeeze himself back into the spacecraft.

Three months later, Edward White made the first American space walk, drifting on the end of an umbilical cable outside a Gemini capsule. Just as Leonov's suit had done, White's also swelled like a balloon, and he too found it hard to move his arms and legs and had difficulty

Commercial companies will play a major role in future lunar projects. ‹

Future lunar explorers will need better space suits than today's. ›

WHERE DID THE MOON COME FROM?

The best theories suggest that the eccentric orbit of a newly formed planet nearly the size of Mars intersected with Earth's more circular path. The two worlds collided about 4.5 billion years ago. Gigantic chunks of molten wreckage scattered into space, and when everything settled down, the result was a remolded Earth and a newly coalesced Moon.

A Z-series prototype suit is put through its paces. The armored joints look clumsy, but are in fact easy to move. ❯

squeezing back into his seat at the end of his walk. At least the Gemini hatch was wide enough to let him back in.

The Apollo lunar space suit did not suffer from the same problems. The upper torso, and the arms and legs, were strengthened with reinforced nylon fabric that did not turn into a balloon when inflated with air; the suit maintained a constant volume. Rubberized bellows at the shoulders, elbows, hips and knees allowed an astronaut to bend his arms and legs without too much difficulty, although they did have to work quite hard against the resistance of the stiff materials.

Nothing much has changed today. When the outer white fabric layers of an ISS suit are stripped away, its armored secrets are revealed like some kind of alien robot. Space suits are far from perfect. Space walkers are frustrated by simplest things, such as the lack of visibility provided by their helmets. The next generation of suits for future lunar and Mars missions will include a larger helmet that provides better visibility. The latest prototypes, the so-called Z series, have been compared with Buzz Lightyear's outfit from the Toy Story movies. In fact, the shoulder and knee joints are more mobile than anything worn by either the fictional astronaut Buzz Lightyear or the actual astronaut Buzz Aldrin. The Z-series can do something that no space suit has ever managed before. It enables the wearer to swivel his or her hips and to bend down, twist, or crouch more or less like an unclothed person.

Space suit gloves are always problematic. They are too stiff to handle delicate tools. Sensitive fingertip control has to be sacrificed, because this is exactly where a suit is most in danger of puncturing through wear and tear. When ISS astronauts peel off their heavy suits after a long session, it is not uncommon for them to reveal bruised or bleeding fingers. With sunrise and sunset alternating every 90 minutes as the ISS speeds around the Earth, a long spacewalk can throw a suit's cooling circuits into confusion. In February 1995, astronauts Michael Foale and Bernard Harris had to be pulled in early from a spacewalk in case they lost their fingers to frostbite. Kathy Thornton,

a veteran of the famous Hubble repair missions, says that working in a space suit is "like trying to fix the carburetor on your truck while wearing thick baseball mittens, and it's like you're skidding about on an ice-rink because you've got no purchase. On TV it looks a breeze, but it's truly exhausting."

On July 16, 2013, space-walking Italian astronaut Luca Parmitano nearly gained the appalling distinction of becoming the first man to drown in orbit. His water-cooling system sprang a leak. Instead of pooling at his feet, the loose water clung to his face, obscuring his vision, and getting into his nostrils and mouth. As ISS swept around the night side of the Earth, Parmitano quite literally had to feel his way around the giant platform, one handhold at a time, until he could find the airlock and get back on board.

Current prototypes for Mars space suits are constructed from smooth hard shells that can be wiped clean of dust as easily as wiping down the smooth paintwork of a car. A rigid exterior also provides extra protection against punctures. The ingenious knee, hip, elbow, and shoulder joints of the Z-series do enable wearers to move quite flexibly, but all these mechanical requirements threaten to make the suits somewhat heavy for working on Mars. The Apollo suit was designed to operate in lunar gravity, which is one-sixth that of the Earth. On Mars, the gravity burden is more than twice as great. The owner of a bulky suit would soon become exhausted with the effort of moving. Martian suits will have to be made from very light, yet strong materials.

Future suits will have to be far tougher than any previous design. At the same time they will have to be comfortable, because the astronauts will be wearing them day after day. The Apollo suits were built to survive lunar surface missions lasting at most two or three days, but Martian suits will need to function for weeks or months on end without springing a leak. Furthermore, the fine orange-red dust that covers much of the Martian surface is chemically corrosive. Every time an astronaut climbs in or out of a suit, there will be a danger of inhaling dust trapped in the folds and connector sections of the complicated

THE BIOSUIT

Perhaps the traditional bulkiness of space suit designs has been an evolutionary mistake. In theory, a constant body garment could be almost exactly the same shape as a human body. Skin-tight materials could make a space suit as sleek and stylish as a ski racer's outfit. If the layer of air between the wearer's skin and the inner layer of the suit fabric is kept as thin as possible, then the ballooning effect that dogged early astronauts can be avoided. At the Massachusetts Institute of Technology (MIT), astronautics professor Dava Newman is working with colleagues and students on an idea known as the BioSuit. "The suits used on ISS and in the space shuttle are technological marvels," Newman says, "but their biggest problem is rigidity. A skintight suit allows greater mobility."

Newman's design may also be safer than current models. "A puncture in a traditional suit threatens sudden decompression because all the air inside might escape. A small breach in the BioSuit could be easily repaired." Only the small patch of skin underneath a puncture zone is compromised by loss of pressure. Air cannot escape from elsewhere because the BioSuit fits snugly around the rest of the body. Adhesive tape could cover a puncture for long enough to allow the wearer to reach safety. This design is lightweight and flexible: no more uncomfortable than a diver's wetsuit. More tests will determine if lunar or Martian dust can be kept at bay.

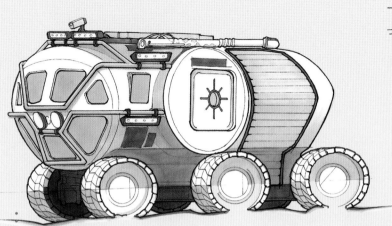

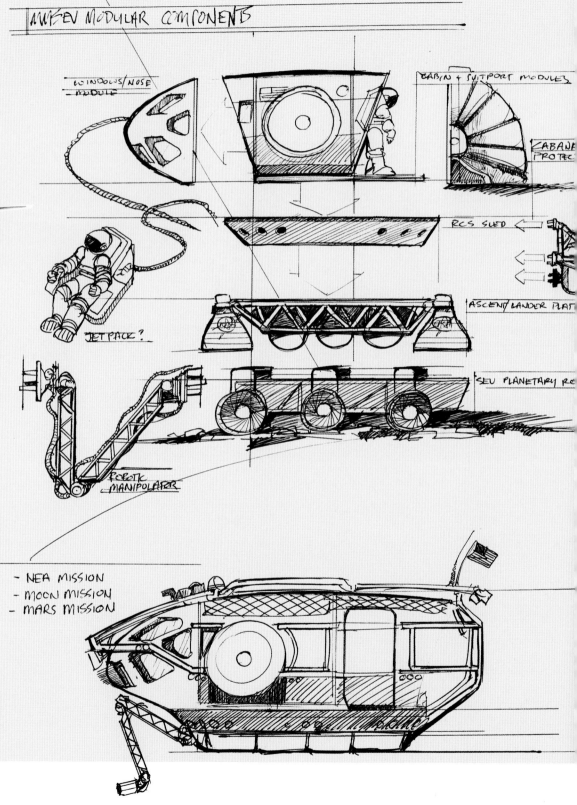

WINDOWS/NOSE MODULE

CABIN + SUITPORT MODULES

CABANE PROTEC

RCS SLED

JET PACK?

ASCENT LANDER PLAT

ROBOTIC MANIPULATOR

SEV PLANETARY R

- NEA MISSION
- MOON MISSION
- MARS MISSION

SHAPE-SHIFTING SPACESHIPS

When is a spaceship not a spaceship? When it can either fly or drive. NASA's concept for a Multi-Mission Space Exploration Vehicle (MMSEV) takes a standardized crew module and attaches it to different support systems, whether a wheeled chassis for planetary surface exploration on the Moon or Mars, or a set of thrusters and a robot grappling system for working alongside an asteroid. The MMSEV system is designed to be flexible, depending on the destination. Prototypes equipped with Z-series space suit ports are already being tested on Earth.

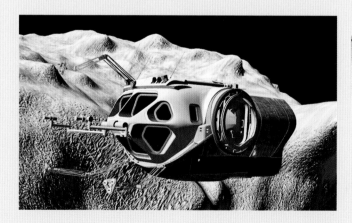

A training prototype for the Multi-Mission Space Exploration Vehicle (MMSEV). ⋀

The MMSEV's large windows give the crew an excellent view of external robot arms and other tools. ⋁

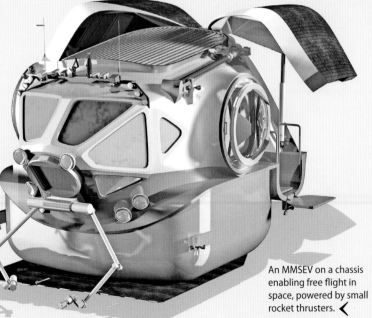

An MMSEV on a chassis enabling free flight in space, powered by small rocket thrusters. ⟨

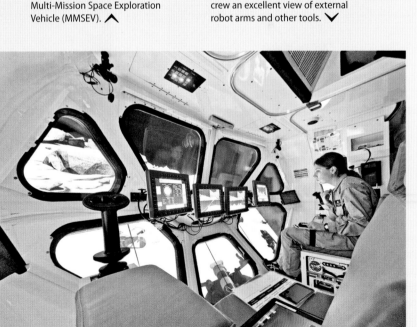

> **"ALL THE NECESSARY CONDITIONS TO PERPETRATE A MURDER ARE MET BY LOCKING TWO MEN IN A CABIN FOR TWO MONTHS"**
>
> Cosmonaut Valery Ryumin, 1980

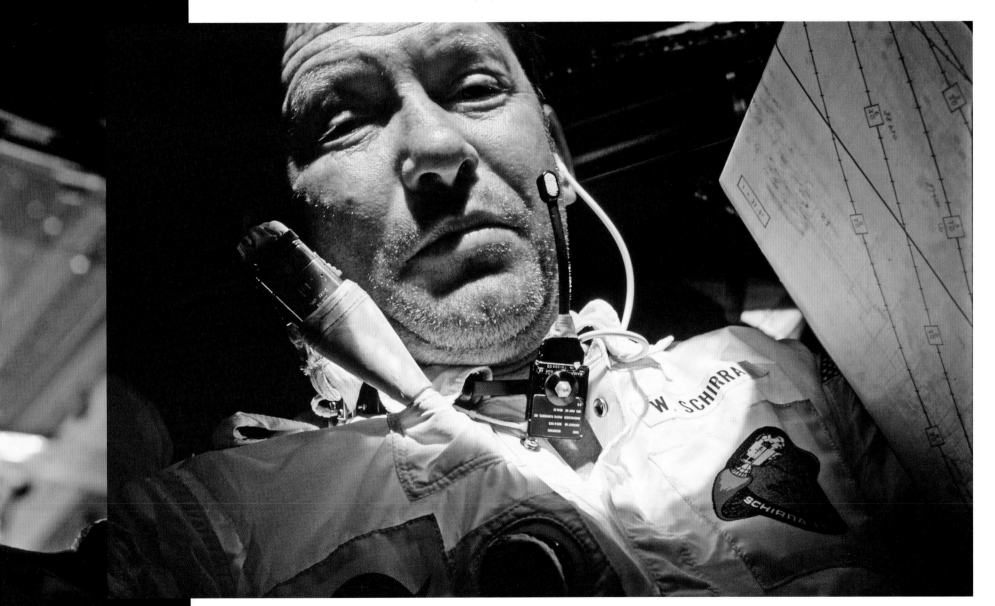

Apollo 7 commander Wally Schirra shows the strain induced by a head cold, exacerbated by clashes with Mission Control, in October 1968. ◀ ▲

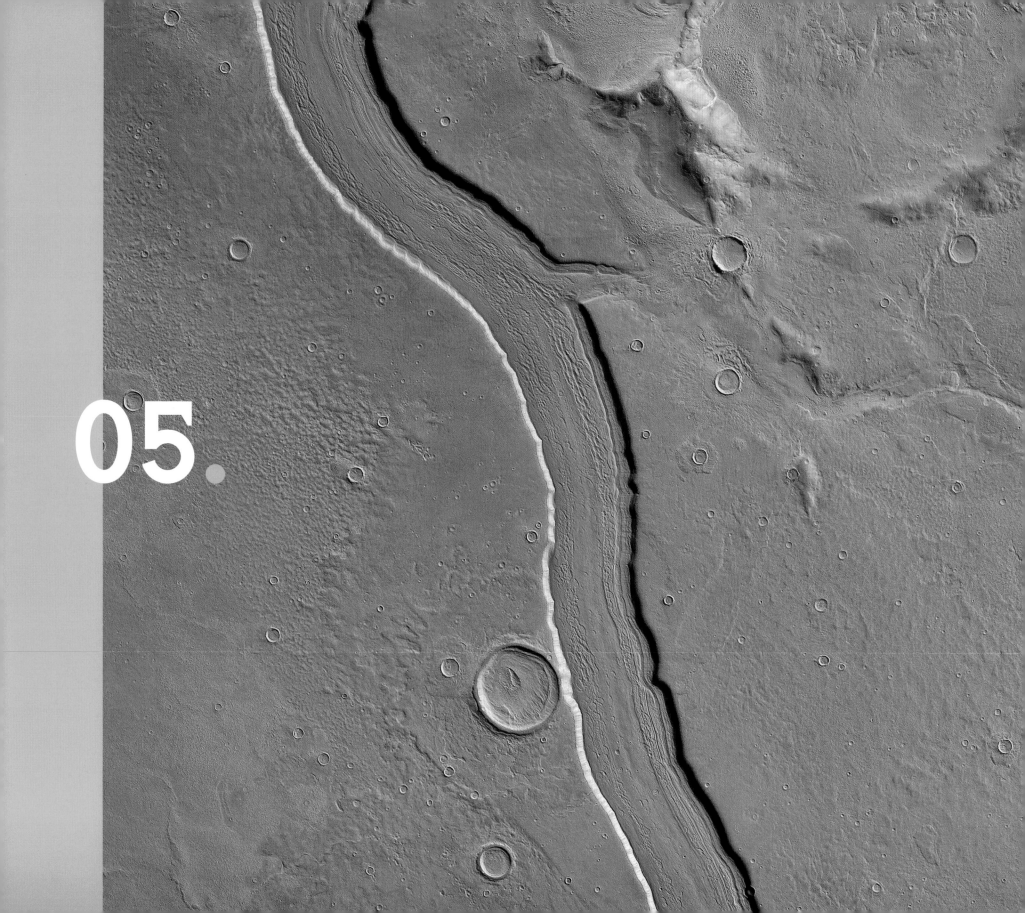

05.

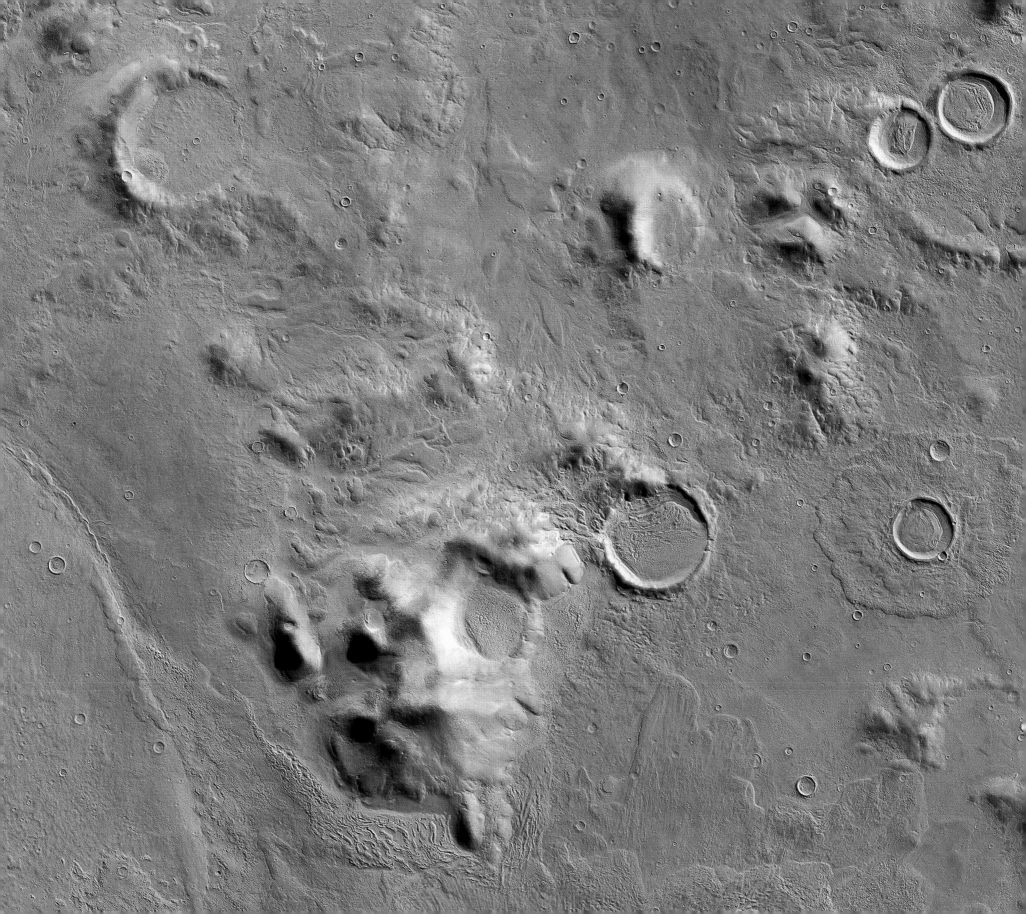

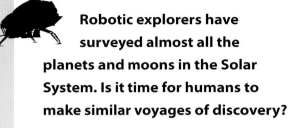

Robotic explorers have surveyed almost all the planets and moons in the Solar System. Is it time for humans to make similar voyages of discovery?

INTERPLANETARY
ADVENTURES

The Mars Pathfinder mission opened a new era of Internet space exploration. As the little wheeled rover *Sojourner* beamed back its startlingly clear images of the planet's rusty red surface, the relevant NASA websites received a world record thirty-three million hits on July 4, 1997, touchdown day. Four days later that figure had risen to forty-seven million. A small Internet team led by Kirk Goodall of the JPL in California had revolutionized public awareness of planetary exploration on a budget so small, it made almost no impact on Pathfinder's already modest costs. We now expect to be able to log on and retrieve data from planetary rovers or deep space probes without a second thought. Early theorists imagined that we would visit the farthest reaches of the solar system. What they did not expect was that those regions would be brought to us, without anyone having to leave the house.

Modern 3D landscape scanning technology makes looking at Mars seem almost as tangible as looking at our backyard. Images of its surface can be downloaded on home computers within a few days or even hours of

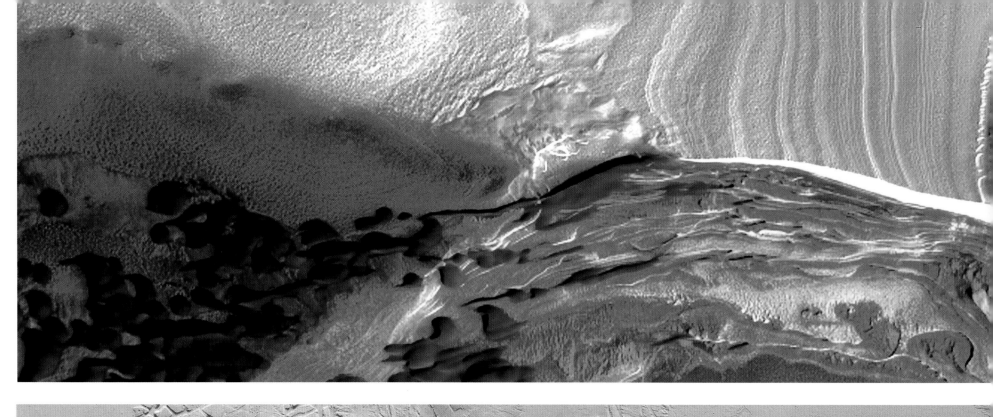
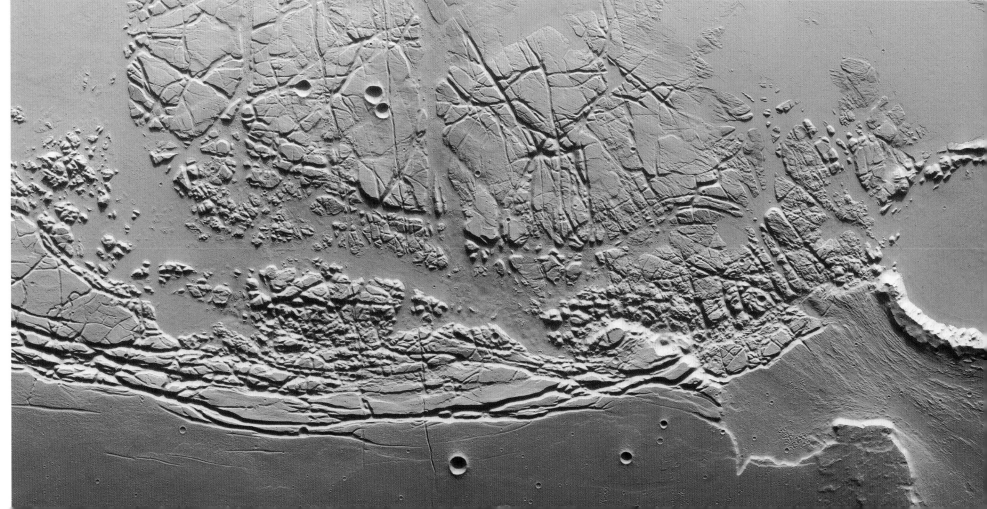

Polar ice on Mars consists primarily of water, capped by a thin layer of frozen carbon dioxide. ❮

Reull Vallis is just one of many features on Mars that appear to have been shaped, in part, by ancient water flows. ❯

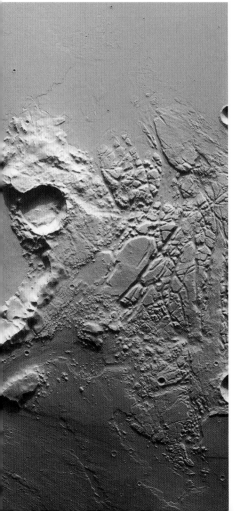

transmission. Ground controllers need time to process the data, but essentially they become public as soon as they are ready. We now take for granted the perspectives we have of Mars, Jupiter, Saturn, dozens of their moons, and even several major asteroids.

Does this mean, then, that we no longer have to go to the trouble of turning up in person to clutch handfuls of alien dust in our human hands? This has been the great debate ever since the space age began. Logic suggests that machine probes are the safest and most cost-efficient tools for space exploration. Instinct and emotion cause many of us to think differently. The world's major space agencies continue to make plans for putting humans on Mars in the coming generation. Today they are joined by a number of determined private entrepreneurs.

It is easy to imagine that NASA's engineers could send humans to Mars within a decade, if only politicians would let them to get on with it. In 1961, at a time when we barely knew how to reach low Earth orbit, John F. Kennedy championed the idea of a manned lunar landing, and the task was accomplished in just eight years. Today, surely we know enough about space systems to make even faster progress across the solar system, so why cannot a modern president give the green light to the red planet? Actually,

some presidents have tried. On July 20, 1989, President George H. W. Bush celebrated *Apollo 11*'s twentieth anniversary. With Armstrong, Aldrin, and Collins standing at his side, he said, "For the1990s, we have the space station. For the new century ahead, we should go back to the Moon." Then, in words that were music to the space community's ears, he talked of "a journey into tomorrow, a journey to another planet, a manned mission to Mars." A NASA team set to work on the now infamous "90-Day Study," so named because this was the time allocated for completing it. Unfortunately, that three-month period was all it took to devise a plan guaranteed to make Bush wish he had never mentioned Mars. A space station was to service a thousand-ton interplanetary craft assembled in Earth orbit. The Mars return trip would take eighteen months, of which barely a fortnight would be spent on the planet's surface: just enough time to plant a flag and snap some photos before heading for home. The price tag was at least $450 billion.

Congress balked at the costs. At the Martin Marietta company, engineer Robert Zubrin was similarly shocked. It seemed that building a giant spaceship had become more important than the mission itself. "I fired off a memo, saying it wasn't enough simply to reach our destination. We had to

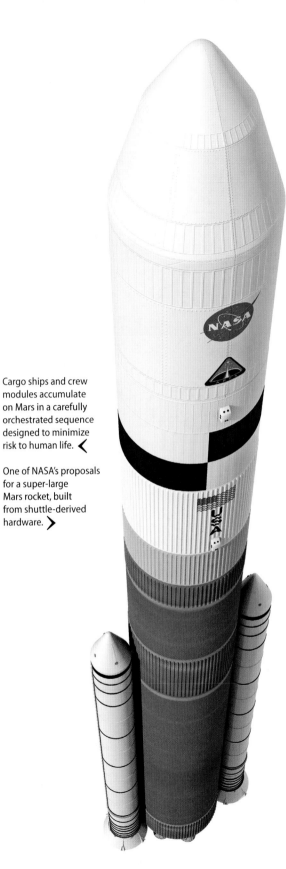

Cargo ships and crew modules accumulate on Mars in a carefully orchestrated sequence designed to minimize risk to human life. ◀

One of NASA's proposals for a super-large Mars rocket, built from shuttle-derived hardware. ❯

A crew module rises from the Martian surface and docks with an Earth return vehicle waiting in orbit. ▲

do something useful when we got there. I thought the plan was totally wrong, and too expensive, and many people at NASA were upset when I spoke out of turn." Some industry colleagues were suspicious of Zubrin's outspoken language, but he felt justified in making his feelings felt: "Aerospace companies usually tell NASA exactly what they want to hear, because that's the way to make a sale. I was proposing to do the opposite and tell the truth, whether NASA liked it or not. Theirs was the worst and most inefficient way to get to Mars."

Mars Direct

Al Schallenmuller, Marietta's chief of civilian space systems at that time, allowed Zubrin to rewrite the company's official sales pitch to NASA. By February 1990, Zubrin's team had reduced the space station's role, cut the weight of the Mars craft in half, and slashed the costs. But Zubrin was not yet satisfied. Time spent on the Martian surface was still only a few weeks at the end of a six-month outward trip for

the crew. The construction of the Mars ship in Earth orbit annoyed him because it cost money and wasted fuel. So he conceived Mars Direct, a scheme in which small vehicles lift off from Earth and head straight for Mars without wasting consumables hanging around in Earth orbit.

Like many an engineer before him, Zubrin began by discarding long-held assumptions. "Most Mars plans call for a huge mothership to circle the planet and send down small landing teams, which then come up, rendezvous with the ship, and fly home. I call it the 'Battlestar Galactica' approach. Why have the mothership at all? In Mars Direct, you fly items of hardware directly to Mars, and then a small Earth return vehicle (ERV) fires off the surface and heads back home." Astronauts are confined to relatively small cabins for the six-month outward and eight-month return trips, "but we know from our space station experience that people can tolerate that if they're sufficiently motivated. We don't have to build giant space cruisers to go to Mars."

Missions are launched only when Earth and Mars

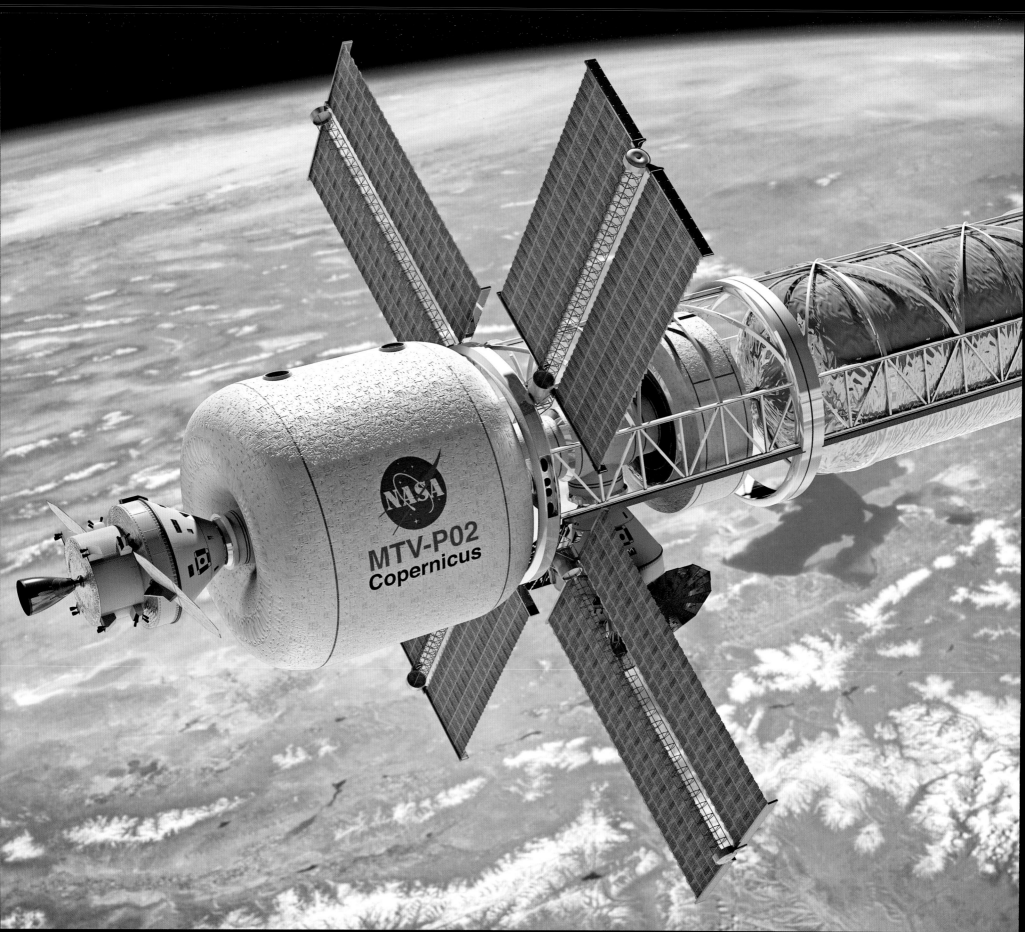

NUCLEAR SPACESHIPS TO MARS

The scientists who built the first atom bombs and nuclear reactors back in the 1940s knew perfectly well how to apply their technologies for space. A design proposal from 1944 suggested that a small nuclear reactor could heat hydrogen fuel to a temperature far beyond those associated with conventional oxidized burning in a chemical rocket engine. Higher temperatures generate higher energy in the exhaust plume, while using less fuel (and no oxidizer). Furthermore, a nuclear engine could apply acceleration to a Mars spacecraft for many hours, while conventional chemical engines run out of thrust in a matter of minutes. We could shorten transit times to Mars from hundreds of days to just three or four months.

Of course, there are risks. The plutonium rods in a nuclear engine's core must first be lifted off the Earth. The brief atmospheric phase of such a launch would be a controversial matter. Although the risks of disaster are low, everyone is wary of the potential consequences. Plutonium dust in the air could cause life-threatening problems for many thousands of people on the ground.

In the 1960s, NASA did develop a working prototype, the Nuclear Engine for Rocket Vehicle Application, or NERVA. Despite a series of extremely encouraging ground tests, the program was cancelled in 1973, mainly as a consequence of budget cuts across the entire spectrum of space activities, but also because of the environmental concerns about nuclear materials. International law limits what can be done in terms of "above ground" nuclear experimentation and forbids the deployment of nuclear reactors in space. These laws might be adaptapted for peaceful international space missions. Development of nuclear rockets has been revived. Although they are not the only way of getting us to Mars, they do promise to be the fastest way.

An Orion is docked to the inflatable habitat module of a Mars transit vehicle in this NASA rendering. ‹

> ## "THE QUESTION OF MARS AS AN INTERPLANETARY GOAL IS NOT SIMPLY ONE OF AEROSPACE ACCOMPLISHMENT, BUT OF REAFFIRMING THE PIONEERING CHARACTER OF OUR SOCIETY"
>
> Robert Zubrin, 1996

swing close to each other in their orbits and are both on the same side of the Sun. This happens roughly every two years. Mars Direct employs a rolling schedule of missions to coincide with close planetary approaches. The downside to this low-energy trajectory scheme is that a manned mission takes more than two years to complete, because the crew has to wait eighteen months on Mars, until the planet is scheduled to swing close by the Earth again, and they can set off on the homeward trip. "In theory this lengthens their exposure to solar and cosmic radiation hazards," Zubrin acknowledges, "but Mars's bulk is a pretty good radiation shield. The mothership method gets you to Mars and back in eighteen months flat, but you're spending most of that time in space, which is where the main radiation hazard comes from."

Zubrin's original scheme took advantage of rocket power already available in NASA's hangars in 1990. The giant liquid fuel tank from a space shuttle was attached to an upper stage with a payload cylinder. At the base, a pod with four shuttle liquid engines was fed from the big tank. Two solid rocket boosters gave extra impetus, just as on a normal shuttle mission. The key point was to leave the winged orbiter component back home and replace its weight with Mars hardware. In theory, this configuration, known as Ares, could have lifted 130 tons of payload into space, just 10 tons less than the giant Saturn V boosters of the Apollo era. But would it have been enough to enable a Mars mission? "We found we could launch a Habitation (HAB) living module and surface equipment, or else an earth return vehicle (ERV), but we couldn't lift everything at once." Zubrin's explorers could fly out to Mars, or else the equipment for their return journey could be dispatched, but no single rocket launch could accomplish all the required tasks.

Far from dooming the plan, this problem turned out to be a great solution, not just in terms of fuel efficiency but also for human safety. Zubrin had a flash of inspiration. "For the first mission, don't even think about sending a crew! Send the ship out empty! You save a huge amount of weight on oxygen, water, food, and other life support, and you lose the weight of the people." It seems crazy, but this approach is what makes Mars Direct so clever. The first component to reach Mars is an empty landing vehicle, with its ERV rocket and capsule installed. Only when this is

sitting safely on the surface are humans risked for the next voyage, knowing that a ride back home is waiting for them.

Once an ERV is safely waiting on Mars, a second rocket lifts off from Earth, carrying inflatable habitats and support equipment, but still no crew, and only a bare minimum of braking rockets and heat shielding for the habitat systems. The touchdown for this kit can be rough and ready, perhaps cushioned by airbags, because the equipment on board is not so critical, or delicate, as the ERV with its essential avionics.

At last, when the planetary alignments are once again appropriate, a third rocket launches the crew. Docked to their Mars transfer ship is a small module capable of landing on Mars, but not of taking off again. There are sufficient supplies on board for the journey out, but not for the return leg, and of course, there is hardly any equipment for Mars surface activities, because most of that has already been delivered by previous launches. You see the general trend. Each outward flight carries only what it must, and no more. A safety contingent might include a small module of emergency rations and an Earth return propulsion stage parked in Mars orbit, just in case the astronauts' landing is aborted at the last moment and they cannot reach the ERV waiting for them on the planet's surface.

Spinning a Line

Flying to Earth from Mars in a small ERV crew module would not be much fun. There are many variations on the Mars Direct scheme, including the possibility that, after lifting off from the Martian surface, each ERV would dock with a small habitat and propulsion module waiting in orbit, thereby giving the astronauts a little more habitable space and propulsive kick for the trip to Earth. At least they would know they were homeward bound, and that help for their exhausted bodies would be immediately available when they reached Earth. However, there would not be any medical assistance waiting on Mars, so they would need to arrive as fit as possible. Zubrin proposes that artificial gravity could be generated, at least for the outward trip. After the upper stage of the launch vehicle has finished firing, its empty tanks will be hurtling through space still attached to the crew module. By disconnecting the rocket stage and swinging the module on the end of a long

The sunlight that helps power robotic Mars rovers could be used to grow food plants for human settlers. ∧

tether, the two components can be made to spin around each other, creating centrifugal gravity inside the crew compartment. When Mars comes within sight, the spinning is stopped, the tether is released, the empty rocket stage is discarded, and full weightlessness is resumed.

The trickiest problem is radiation and its potential effects on the health of the crew. Mars enthusiasts point out that radiation does not kill people. What it actually does is increase the risk of radiation-related cancers in later life. So far, the overall death rate of astronauts shows that (accidents aside) their longevity is no different from

normal. Even so, NASA does not wish to send humans to Mars until the radiation problem has been considered in greater depth. New types of shielding certainly will help. Traditionally, spacecraft shells have been built from aluminum and other metals. Modern plastic composites provide much better protection against radiation. Storing many months' worth of frozen food on the walls of the habitation modules could boost their shielding without increasing the spacecraft's weight. There is even a proposal to store human digestive waste in the wall cavities. The "solids" are quite good radiation shields.

During a simulated NASA Mars mission, volunteers are not allowed to leave their living quarters unless wearing space suits. ∧

MARS ON EARTH

Devon Island in the arctic Nunavut Territory of Northern Canada has some surprising features, not least of which is the fourteen-mile-wide wide Haughton Crater, gouged out 23 million years ago by a meteorite impact. In 1997 NASA scientist Pascal Lee set up a project to explore the crater and its environs. When he saw it for the first time, he knew he had found a perfect place for training future Mars explorers. He contacted the Mars Society, the five-thousand-strong group of scientists, engineers, and space enthusiasts dedicated to putting humans on Mars as soon as possible. After raising over $1 million in sponsorship from the Discovery TV channel and the Flashline software company, the Society built a replica Mars Habitat on Devon Island. It is freezing cold here, but also very dry. Pascal Lee believes, "This is as close to being on Mars as we can get without actually leaving the Earth."

With support from NASA, the Mars Society borrowed a U.S. Marines C-130 transport plane to fly all their equipment to the island. There was nowhere in the rough landscape for the huge plane to touch down. Instead, everything was packed into crates and dropped by parachute. One load slipped out of its harness and smashed to pieces when it hit the ground. Thousands of miles from help, and in bitterly cold conditions, the team had to improvise a way of finishing their base. It was a valuable rehearsal for a real Mars mission. A second "Mars Analog" base was established in the the rocky canyons of southern Utah.

The Mars Society and its partners run regular simulated missions on Earth, testing human endurance. ◄

Living off the Land

One seemingly insane aspect of Zubrin's proposal dates from its first design iteration in 1990. It has been proven to work, at least in a terrestrial laboratory. The ERV module is sent to Mars with very little fuel on board to lift it off again, let alone push it out of Mars's gravity field and toward home. In-Situ resource utilization, or ISRU, derives most of what's needed from the Martian atmosphere. As with all the various Mars Direct elements, this cuts the weight of each piece of hardware that has to be lifted off the Earth. Zubrin calls ISRU "living off the land." In fact, the chemical process for this dates back to the Victorian era. "We didn't have to invent anything new."

Mars's atmosphere consists mainly of carbon dioxide. The ISRU plant pumps this through a nickel catalyst, adding a trace of hydrogen into the mixing chamber. The catalyst splits the carbon dioxide, liberates the oxygen, and combines it with the hydrogen, making water. The freed carbon reacts with spare hydrogen to create methane, which is pumped straight into the ERV's fuel tanks. Meanwhile, a weak electric current passes through the water to extract the oxygen, while the water's hydrogen, now free again, is pumped back into the system so that the cycle can begin anew. This is a simple electrolysis process, familiar to all of us from science lessons. (Some drinking water may be retained, but in the main, astronauts will extract water directly from Martian surface ice and by recycling their urine and exhaled water vapor.)

Six tons of hydrogen carried to Mars and fed into an ISRU could deliver twenty-four tons of methane and forty-eight tons of oxygen—sufficient for an ERV's liftoff requirements plus a modest surplus to help power surface equipment and a rover vehicle. ISRU is powered by the same kind of miniature plutonium thermoelectric generator that has already been used for the Galileo mission to Jupiter and the *Cassini-Huygens* voyage to the Saturnian system. Despite environmental concerns, these units are safe to launch, because the plutonium is contained in exceptionally tough casings. No new technology has to be invented to make ISRU work.

Marketing Mars

In November 2012, SpaceX founder Elon Musk outlined his exploratory vision before a packed audience at the Royal Aeronautical Society in London. "I'm hopeful that the first human mission to Mars is actually some collaboration of private industry and government," he said, "but I think we need to be prepared for the possibility that it has to be just commercial." SpaceX is already designing an extremely large rocket, called Mars Colonial Transporter, with the aim of establishing a colony on the red planet. Designing the craft is one thing. Raising the cash is quite another. Interim

ONE-WAY TO MARS?

Mars One is a non-profit organization based in the Netherlands, led by Dutch entrepreneurs Bas Lansdorp and Arno Wielders. Their mission plan calls for humans to reach Mars and not come back. No provision will be made for a return to Earth. It sounds reckless, but over 10,000 people expressed an interest in becoming one of the four crew envisioned for such a flight. The mission would be funded in large part by global media deals.

Only time will tell if such an unusual concept can work, or even if it would be permitted by relevant nations and governments. Megan Gannon, news editor for the (indispensable) space.com website, sums up the drawbacks of the scheme terrifyingly well: "They'll live out their days in a thick-walled habitat, protected against harmful solar particles and cosmic rays, donning space suits to go outside in a place that lacks a breathable atmosphere."

projects may involve robotic sample return payloads touching down inside Red Dragon, a variant on SpaceX's current capsule technology. A miniature ERV would lift off with precious samples. However, Musk is determined not to be limited forever just to robotic payloads. New and innovative funding techniques will have to be applied to future missions.

Going to Mars and Not Getting There

Inspiration Mars was created by Dennis Tito, the man who opened the private space frontier by flying to ISS in 2001 aboard a Russian Soyuz spacecraft funded by his own cash. His latest scheme calls for a fully qualified pair of astronauts (ideally, a happily married professional couple) to fly around Mars without attempting a landing. Tito is an experienced technologist and knows that touchdown and surface survival systems for human explorers are not yet available to us. On the other hand, the equipment for a much simpler flyby mission is already operational. We know how to launch modules, life-support systems, and other hardware capable of sustaining human life in space for many months at a time. We know how to get off the ground and out of Earth orbit. Above all, we know how to get home again from deep space, using capsules that can withstand the punishing reentry speeds. Additionally,

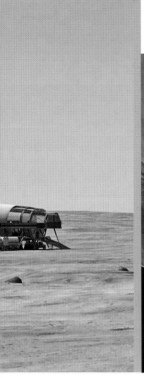

NASA is developing inflatable heat shields for future Mars-bound heavy payloads. ‹

Elon Musk's SpaceX company aims to land adapted Dragon capsules on Mars. ≪

Inspiration Mars is fairly realistic on the funding front. There are no Battlestar Galactica-inspired ships here. Rather, Tito has set aside $100 million for initial development while recognizing that government help will be essential to get the mission off the ground. If NASA can provide a Space Launch System rocket and an Orion capsule, Inspiration Mars could fund other hardware, such as a Cygnus cargo module from Orbital Sciences, adapted as a living compartment for the two astronauts. This just might be the foundation for the first human interplanetary voyage.

Although planetary alignments make the year 2018 a particularly attractive window of opportunity, it is unlikely that the hardware for a flyby could be funded and constructed by then. Similar opportunities will emerge during 2021 and 2033. The plan stays on the table as a tantalizingly plausible idea for our generation, marred only by the fact that the two explorers would have barely forty hours in close proximity to Mars before swinging around the planet and heading back toward Earth. On the plus side, the flight plan guarantees that their little ship will be sent homeward by Mars's gravitational field. No dangerous rocket burns are required for the homeward leg of the 501-day round trip. One trajectory map (for a 2021 mission) uses the gravitational field of Venus to accelerate the spacecraft on toward Mars, allowing flybys of two planets for the price

of one, so long as the crew is ready to tolerate a 588-day flight. Tito's written testimony to the U.S. Congress, dated November 20, 2013, certainly caught lawmakers' attention, and it continues to spur debate. Inspiration Mars "is not motivated by business desires, but to inspire Americans in a bold adventure in space that reinvigorates U.S. space exploration," he said. "It is a philanthropic partnership with government to augment resources and achieve even greater goals than is possible otherwise. . . . It would attract significant private funding, while enabling NASA to do what it does best, and confirm the United States as the unquestioned leader in space." Whatever shape the technical and financial arguments adopt over the next few years, there is certainly an appetite for sending humans to Mars—and an increasing impatience with our failure to send people any further than the Moon.

The Red Planet

Mars has some of the most dramatic terrain in our solar system, including four colossal volcanoes. The greatest of them, suitably christened Olympus Mons, covers an area equivalent to Arizona and is capped by a crater that could swallow the entire island of Hawaii. Another prominent feature, Valles Marineris, is a tangled network of wide, deep canyons wrapping its way around half the planet. These

> " TO TAKE PART IN NEW DISCOVERIES IN THE COSMOS. TO ENGAGE IN A SINGLE-HANDED DUEL WITH NATURE... COULD ANYONE DREAM OF MORE?"
>
> Yuri Gagarin, 1961

Inspiration Mars exploits the lifting power of NASA's Space Launch System. ❮

1 Insertion into Earth orbit
2 Docking with the habitat
3 Escaping Earth's gravity
4 Coasting in deep space
5 Swinging past Venus
6 Flying around Mars
7 Heading back to Earth ❯❯

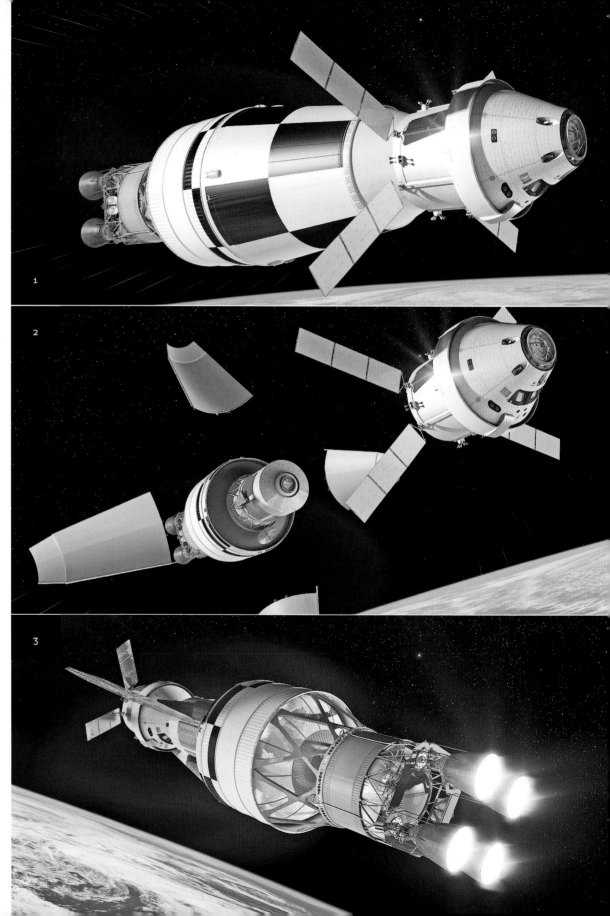

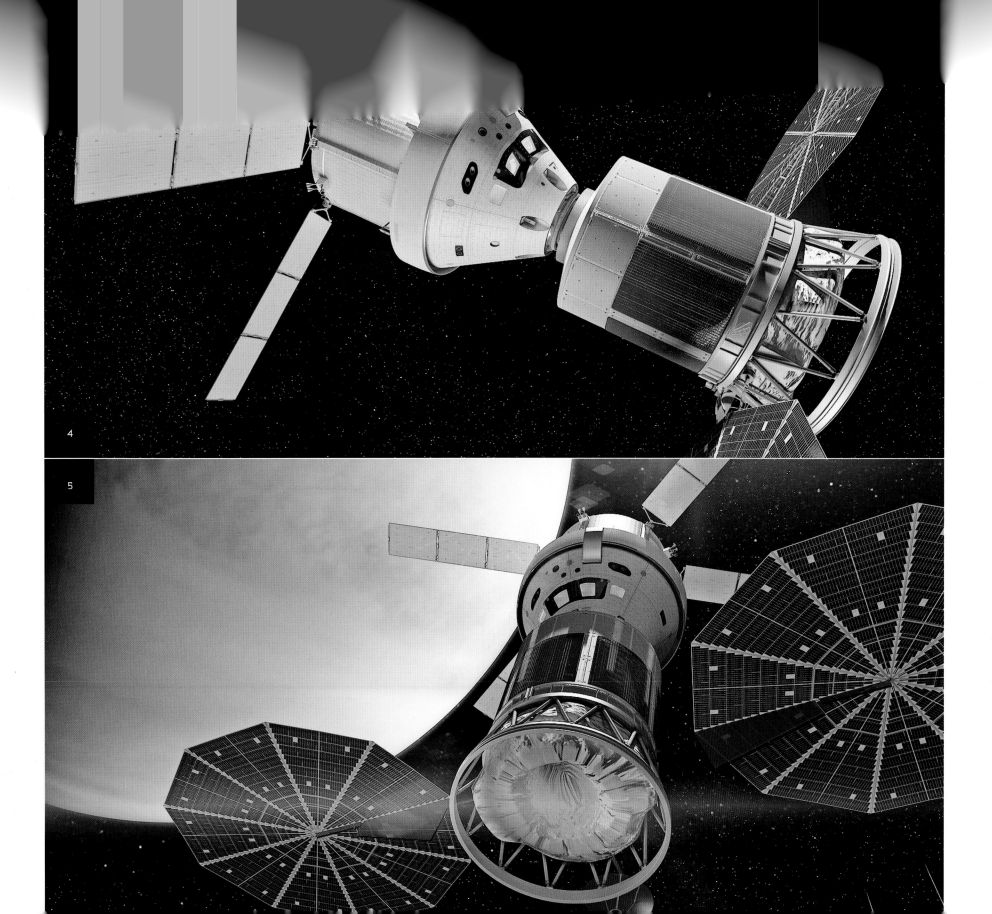

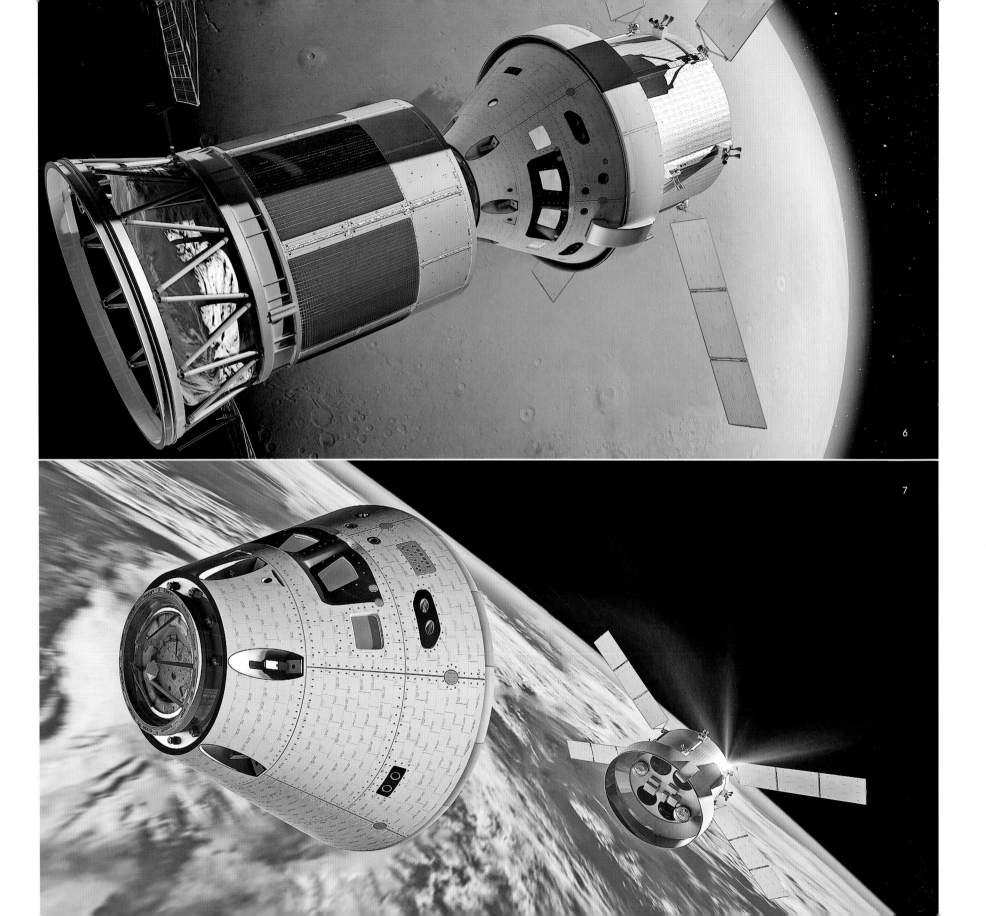

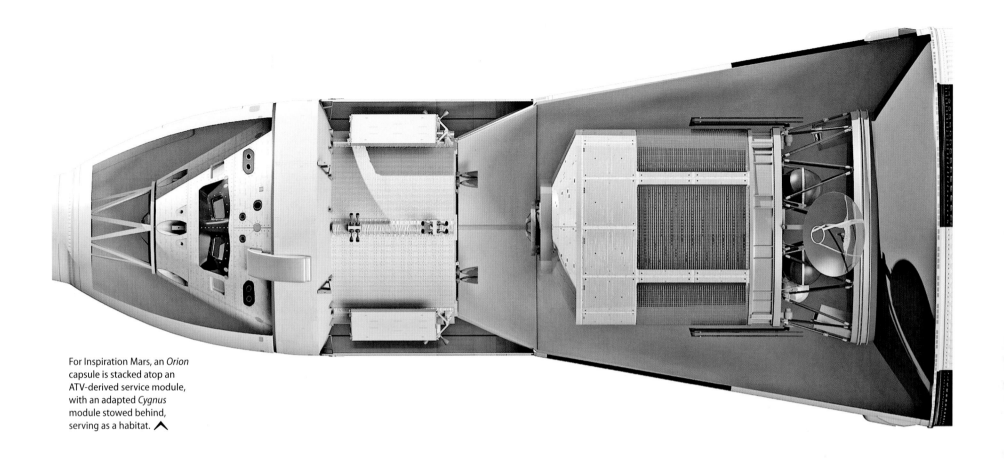

For Inspiration Mars, an *Orion* capsule is stacked atop an ATV-derived service module, with an adapted *Cygnus* module stowed behind, serving as a habitat. ⋀

landscapes are so large, they can be appreciated properly only from space. The largest feature is the least obvious. The major volcanoes, the Marineris trenches, and large areas of rifts and fractures are associated with the Tharsis Bulge, a region where the planet's sphere was pushed out of shape by the internal pressures of molten rock beneath the crust. The volcanoes were fed by this great blister's hot interior.

Images acquired from orbit show ancient riverbeds and sedimentary soils. Scientists still hope to find some microbial life underground, because Mars did once have a much denser and warmer atmosphere than it has today, along with plenty of water. Unfortunately, over billions of years, most of the atmosphere has escaped into space. Though substantial, the planet's gravity is not quite strong enough to hold the gaseous mantle in place. Much of it has escaped into space, leaving a dry and frozen world soaked

in harmful ultraviolet radiation from the Sun. The volcanoes, too, are no longer erupting.

While we will never find beautiful green-skinned Martian princesses on this frozen world, but we might find fossilized evidence of past microbial life. This would certainly rank as one of the greatest discoveries of the millennium. Perhaps some initial clues are already with us here on Earth. Somewhere around fifteen million years ago, a fist-sized chunk of rock was blasted from the Martian surface: collateral damage from a meteorite collision. After drifting through space for fourteen million years and more, this wayward fragment of Mars encountered the Earth and fell onto Antarctica, where it lay hidden under the ice for thirteen thousand years until its discovery by an American meteorite hunting team in 1984. (Meteorites can fall onto any area of the Earth, but they are easier to identify on

pristine polar ice sheets, where terrestrial rocks are rare.)

In the summer of 1996, a team of scientists led by NASA astrobiologist David S. McKay announced the discovery of what they believed were fossilized traces of life in the meteorite, known as ALH84001. President Bill Clinton said at the time, "If this discovery is confirmed, it will surely be one of the most stunning insights into our universe that science has ever uncovered." Unfortunately, we still cannot prove that the meteorite's tantalizing mineral traces definitely owe their origins to ancient Martian bugs, and skeptics cannot prove their case beyond doubt, either. This problem can best be solved by investigating rocks on Mars itself. Meanwhile, stray space rocks of all kinds, whether from Mars or not, are worth studying—not just because of their scientific fascination, but just in case one of them threatens to destroy us all.

In this artwork from NASA, astronauts from a docked Orion inspect the capture bag around a small asteroid, retrieved earlier by a robotic craft. ∧

Asteroids: The Wreckage of Creation

Five billion years ago, a vast cloud of dust and gas drifted through our galaxy. These thin, widely scattered wisps originated from supernovae, the explosive deaths of stars from an earlier generation. The cloud, called a nebula, was gently pulled around by gravitational forces from other, surviving stars in the surrounding galaxy. Most of the atoms in the nebula were hydrogen or helium, but all the other chemical elements were present in smaller amounts.

Somewhere within the nebula, a seemingly insignificant clump of matter coalesced, gaining mass until it was capable of exerting its own gravitational force—extremely weak, yet just strong enough to attract nearby atoms and molecules. This mass, or protostar, grew to the point where its gravity became strong enough to pull in more material from ever greater distances: tens, then thousands, and eventually many millions of miles. Dust and gas from the surrounding nebula became concentrated into a swirling, plate-shaped structure known as an accretion disk, with the protostar at its center. This was the beginning of our Sun. As it gathered more and more mass, some of the chemical debris in the accretion disk began to cluster in miniature, and much cooler, versions of the protostar process. Small worlds, known as planetesimals, began to form. When these collided with each other and their wreckage recombined, larger structures emerged and consolidated into the planets and moons that exist today.

One of those worlds became our Earth. It began life around 4.6 billion years ago as a hot, chaotic sphere of molten metals and silicates, but over time the various materials began to divide according to density, with the heaviest metals, such as iron, sinking into the Earth's core, and the lighter materials, such as silica, floating upward toward the planet's surface. After approximately one hundred million years, the outer surface cooled sufficiently for a thin rocky crust to form.

More worlds than we see today condensed out of our solar system's accretion disk, but at least a few of them may have developed orbits around the Sun that were too elliptical for safety. At some point in the distant past, their paths intersected and they smashed together, creating a swarm of rubble. Each chunk is known as an asteroid. Some asteroids may be the remains of an ancient planetesimal that was unlucky enough to stray too close to Jupiter. Tidal forces from the mighty gas giant tore the smaller world (or worlds) apart. The precise history of asteroids is still a mystery. What we do know is that the Earth has been smashed into by rogue asteroids many times, and these crashes are bound to happen many times again in the future. Meteorites are stray asteroids that collide with the Earth, survive their fiery descent and reach the ground more or less intact. Every day the Earth is bombarded by upwards of one hundred tons of small rocks and dust fragments. At least a thousand asteroids a third of a mile and more in size regularly sweep across Earth's orbit. Once every one hundred thousand years or so, something big hits the planet. There are countless rogue rocks out there that we haven't spotted yet. If one the size of a greyhound bus happened to hit New York, the city would be destroyed. An impact from anything bigger than a mile across could wipe out human civilization.

In 1908, a 1,200-foot-wide meteorite blew up in the air above the remote Siberian forest region of Tunguska, releasing energy equivalent to the blast of several atomic bombs. Hundreds of square miles of trees were broken like matchsticks. There were only a few dozen eyewitnesses: loggers and peasants whose reports were not taken seriously by the authorities in Moscow. A full scientific survey of the area was conducted only after World War I. Today we have a much fresher memory to draw upon. On February 15, 2013, an asteroid approximately sixty feet wide, traveling at twelve miles per second, slammed into the atmosphere above Chelyabinsk, Russia. A fireball far brighter than the Sun streaked across the sky. Shock waves shattered glass in thousands of buildings, causing many injuries. People watching outdoors were knocked off their feet by the air blast. Approximately twenty-eight miles above the ground, the asteroid exploded, releasing as much energy as an atomic bomb. Some witnesses were mildly scorched and temporarily blinded.

In witnessing such an event, we get a taste of what an asteroid impact can do. As far as we can tell, the worst relatively recent smash happened approximately 65.5 million years ago, when an asteroid six miles wide struck the Earth. The impact gauged a crater more than one hundred miles across. Erosion has smoothed away many of the sharp edges, but this ancient shock best explains the circular rock density variations overlapping the Yucatan Peninsula and the Gulf of Mexico, a formation now known as the Chicxulub crater. The shock waves alone, in both the atmosphere and the oceans, must have caused

It's happened before and it will happen again: a large asteroid strikes the Earth, creating a massive blast wave of destruction. ∧

catastrophic damage across vast swathes of the Earth's surface. Dust hurled high into the atmosphere obscured the Sun for many months, and possibly even decades, before dissipating. Precious sunlight was dimmed for a longer time than even the hardiest plants and animals could tolerate. The effect on food chains was appalling, especially for dinosaurs, the comparatively large and hungry creatures at the top of the chain.

NASA and other space agencies are in the process of refining our asteroid monitoring systems. But what should we do if we find a large and dangerous rock heading toward us? The renowned planetary scientist Carl Sagan warned that the skills needed to deflect it away from Earth might just as easily be used to create deliberate chaos. Rogue nations or space-based terrorists could blackmail the entire planet. In his last book, *The Pale Blue Dot*, Sagan wrote, "I keep hearing that only a madman would do something like that. I have to remind myself that madmen really exist." Failing that, orbiting entrepreneurs could destroy us by accident, seeking riches but harvesting destruction instead. Some asteroids are rich in platinum,

one of the most precious metals. Every scoopful of drilled material would yield a profit if only the world's mining companies could get hold of an asteroid in the first place.

Blowing up a rogue asteroid would create a swarm of debris almost as dangerous as the original larger mass. A more subtle approach would be to create a nuclear burst alongside an asteroid, deflecting it from its course rather than smashing it. At present, we do not possess missiles capable of reaching sufficiently far into space to catch up with an asteroid. Even if we did fire one, a missiles' journey time might still be measured in years rather than months or days, depending on the distance and trajectory of the target. We would need substantial advance warning of an asteroid hazard, plus plenty of time for the global community to discuss what to do, and who should be in charge. At the moment, placing a nuclear warhead above the Earth's atmosphere is forbidden by international law.

Perhaps the best way to deal with asteroids would be to leave the nuclear weapons in their silos. Instead, we could send simple robot probes to push an asteroid very gently off its doom-laden course. Solar panels could

NASA'S ASTEROID INITIATIVE

NASA has proposed a mission to capture a small asteroid and bring it within reach of us for closer examination. In theory, the propulsion and energy requirements for such a mission are feasible. Every year, several dozen asteroids in the twenty- to forty-foot size range pass closer to Earth than the Moon. Given enough warning, a robotic thruster system could be latched onto one of these, diverting it into a safe orbit around the Moon. The new Orion spacecraft is intended for flexible deep space missions of this kind, and it may be that a gloved human hand will touch the flank of an asteroid long before a booted foot makes an imprint on the surface of Mars.

NASA is investigating how to send a robotic electric-ion thruster to meet a near-Earth asteroid and divert it into a stable lunar orbit, within reach of an Orion spacecraft. The robot capture device may even deploy a bag around the target asteroid, preserving its scientifically valuable accumulation of surface soil and rubble against any disturbances. Some asteroids are very solid lumps. Others are loose clusters of ancient debris. Among suitable candidates, the slow-moving Asteroid 2009 BD may be a good first target. A probe launched in 2019 could rendezvous with it and tow it into lunar orbit by the year 2023.

Studies suggest that small asteroids often are captured by the Earth's gravitational field. Some of them stay within reach for weeks, or even months, orbiting within potential range of spacecraft before finally escaping into deeper space. These "minimoons" are typically only a few feet wide, but scientifically they are almost as interesting as their larger cousins. They could make NASA's mission planning much easier.

NO HAZARD	0	The likelihood of collision is effectively zero, as applies to small objects such as meteors that burn up harmlessy in the atmosphere as well as infrequent small meteorites that cause no serious damage.
NORMAL	1	A large near-Earth asteroid poses no unusual level of danger and creates very little public concern. Regular telescopic observations very likely will drop the score down to Level 0.
MERITING ATTENTION BY ASTRONOMERS	2	An object making a somewhat close pass near the Earth merits close attention from astonomers but without causing undue public alarm.
	3	This threat level applies to objects that pose a 1 percent or greater risk of striking the Earth in less than a decade. Careful monitoring is required.
	4	Any object threatening a 1 percent or greater chance of destructive collision must be watched if less than a decade away.
THREATENING	5	A close encounter less than a decade away, posing a serious, but still uncertain threat of regional devastation, justifies governmental contingency planning.
	6	A very large object posing a serious but still uncertain threat of a global catastrophe requires contingency planning if the encounter is less than three decades away.
	7	A very close encounter by a large object, expected within a century, and posing an unprecedented threat of a global catastrophe, requires international contingency planning.
CERTAIN COLLISIONS	8	A collision is certain, capable of causing localized destruction on land or possibly a tsunami if offshore. Evacuation measures must be in place.
	9	A collision threatens unprecedented devastation. Such events occur on average between once per 10,000 years and once per 100,000 years.
	10	A collision threatens a climatic catastrophe that endangers the future of civilization as we know it.

THE TORINO SCALE
Assessing asteroid/comet impact predictions. ❮

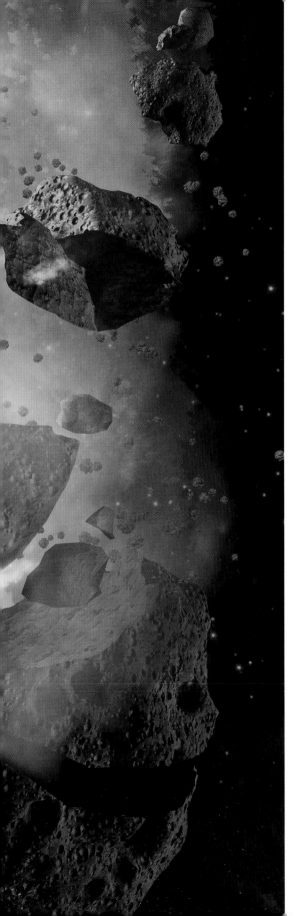

Origins Spectral Interpretation Resource Identification Security Regolith Explorer (OSIRIS-REx) will rendezvous with near-Earth asteroid 1999 RQ36. Approaching to within three miles of the asteroid, the spacecraft will conduct a six-month mapping survey. Mission controllers at JPL will select a location from where a sample can be captured. The spacecraft will move closer, and its arm will extend, gathering more than two ounces of material for return to Earth in the year 2023. RQ36 is approximately two thousand feet in diameter: roughly the size of six football fields. Current telescope evidence suggests that this ancient specimen is rich in carbon and may contain organic chemicals.

Useful asteroids could be steered in our direction just as easily as we could keep them away from us, simply by attaching those ion thrusters (and without any great sense of urgency). After a matter of years, suitable candidates could be delivered into the gravitational influence of the Earth-Moon system, within reach of astronaut investigators. There is even a scheme for aerobraking incoming asteroids in the same way that probes are routinely skimmed through the atmosphere of Mars to slow them down. This technique would place an asteroid into a circular Earth orbit, from where it could be accessed just as easily as we now reach a space station. But the slightest miscalculation could be disastrous. It would be a terrible irony if the next asteroid to hit us were brought here deliberately.

The Search for Life

While Mars might be haunted by the faint mineral ghosts of a long-dead biology, another world in the solar system could be the home of creatures that are very much alive. In August 1996 the Galileo space probe beamed back detailed images of Europa, Jupiter's fourth-largest moon. The white, cold surface material here seems tantalizingly familiar. On Earth, jagged sheets of polar pack ice drift loosely in the sea during summer months, then freeze into place during winter. The surrounding waters also harden, leaving distinct grooves between the thicker ice sheets. Close-up views of Europa's crustal grooves look just the same, but with an added ingredient. It seems as though a dark, slushy deposit occasionally oozes up through the cracks between the floes, leaving murky stains near the surface when the grooves refreeze. This slush seems to emanate from a

subsurface liquid material. Tidal forces from Jupiter keep Europa's surface ice in turmoil, and there is strong evidence that those same forces also prevent the fluid underneath from freezing solid. It may even be slightly warm. In 2014 Hubble Space Telescope data provided strong evidence for giant water plumes bursting out from Europa.

Could this indicate Europa's suitability for life? Even without sunlight, organisms in the hidden ocean could be sustained by heat from hydrothermal vents on the ocean floors. We know this is possible because we've found many such organisms here on Earth. In the late 1970s, scientists in the deep-sea submarine Alvin, operated by the Woods Hole Oceanographic Institution in Massachusetts, plunged to the bottom of the eastern Pacific, some six thousand feet below sea level. They found hydrothermal vents, eerie chimney-like structures belching superhot, mineral-rich fluids into the cold, dark waters above. So-called black smokers typically form along ridges where tectonic plates diverge and new molten crust is pushed from below into the ragged space that opens up. The smokers form when dissolved metals suddenly precipitate as the hot water bursts upwards and meets the surrounding ocean water, which is only a few degrees above freezing. These eerie towers spew a noxious brew of chemicals, giving the dense "smoke" its characteristic dark hue.

Black smokers seethe with extraordinary, unexpected life, such as tubeworms, clams, and eyeless shrimp. While our familiar surface food chain is based on energy from the Sun (grass makes cows, and cows make hamburgers), those rays never reach the deepest sea floors. Here, organisms must rely on a different energy source: the metals and other chemicals that rocket out of those vents. In one especially intriguing reaction, hydrogen sulfide and iron monosulfide deliver the energy for microbial life to thrive. They react to form the mineral iron pyrite ("fool's gold") and hydrogen gas. It's hydrogen, rather than sunlight or oxygen, that provides the energy for the bacteria that are the bottom of the food chain at the bottom of the deepest oceans. Perhaps life on Earth began in these sunless depths? If so, we must widen yet further our ideas about what kind of extraterrestrial environments might make suitable domains for life—not just on our world, but on other planets and moons, too. Certainly Europa is a priority for future space probe planners. Another satellite moon

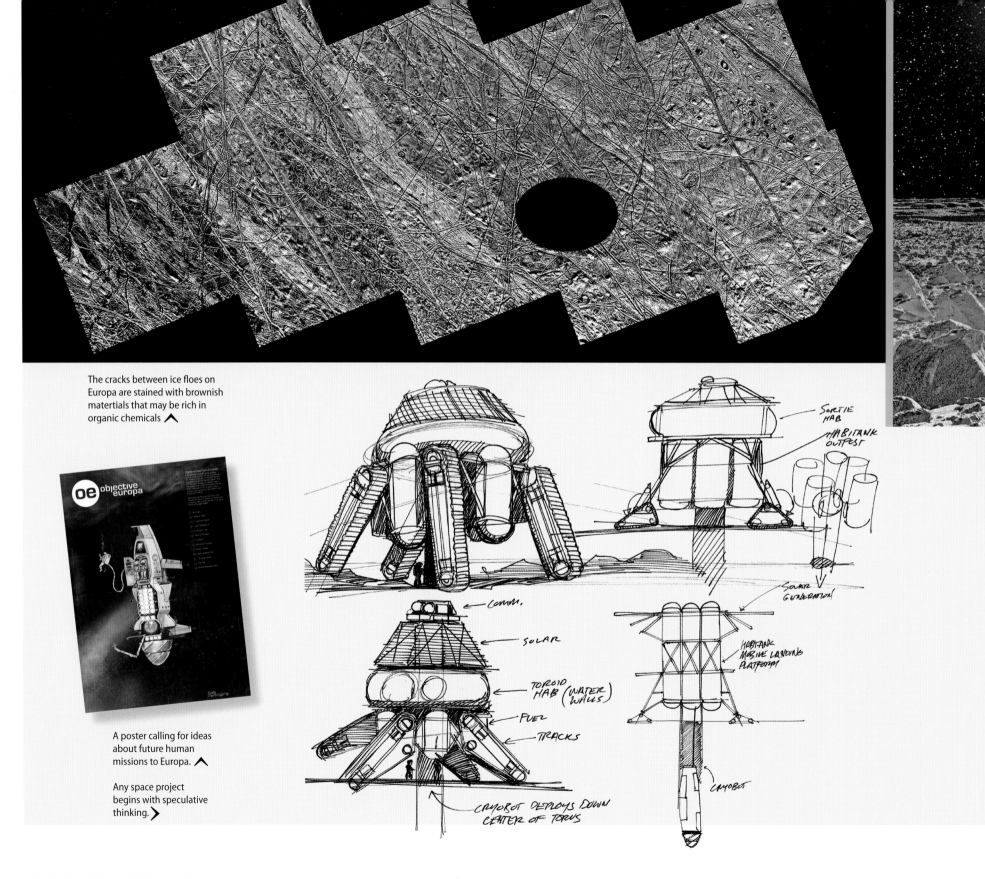

The cracks between ice floes on Europa are stained with brownish matertials that may be rich in organic chemicals ⋀

A poster calling for ideas about future human missions to Europa. ⋀

Any space project begins with speculative thinking. ❯

SORTIE HAB

HABITANK OUTPOST

SOLAR GENERATION

COMM.

SOLAR

TOROID HAB (WATER WALLS)

FUEL

TRACKS

CRYOBOT DEPLOYS DOWN CENTER OF TORUS

HABITANK MOBILE LANDING PLATFORM

CRYOBOT

is generating similar excitement. In 2014, researchers at Sapienza University in Rome published their analysis of data from the *Cassini* space probe. The Saturnian moon Enceladus also harbors a warm ocean beneath its icy crust, and this ocean almost certainly has a rocky floor, where black smokers may be as common as they are on the beds of deep terrestrial oceans. *Cassini* has detected simple organic compounds in the plumes of water vapor ejected by this tantalizing moon.

Even if we do not find life elsewhere in our solar system, there are great prizes to be won from the study of prebiological chemistry. NASA may have been first to land on the Moon and Mars, but on January 14, 2005, a European space probe made history by touching down on Titan, a far stranger and more distant world. *Huygens* took seven years to get there. It spent most of that time clinging silently to the flanks of NASA's *Cassini* mothership as it swept across our solar system on an epic two-billion-mile journey to explore Saturn and its moons. On arrival, *Huygens* was designed to operate only a few hours before losing contact with its mothership. That was enough time to give us a glimpse of an awe-inspiring alien world.

As *Huygens* plunged through the atmosphere of Titan, it confirmed what scientists have long suspected. This is the only satellite world in our solar system with a substantial atmosphere. It consists mostly of nitrogen, but there is also a thick smog of methane clouds. When the cameras switched on at one hundred miles altitude, some real surprises were in store. ESA scientists at Darmstadt in Germany knew from the first frames they had struck gold. The cameras showed that methane rain falls on the ground, and liquid methane rivers scour Titan's landscape as surely as rain carves the landscapes of Earth. From further out in space (using *Cassini*'s infrared camera) regions of light and dark texture were revealed beneath Titan's dense atmospheric haze. *Huygens*'s closer inspection showed the brighter areas are rough terrain, probably consisting of water ice. Dark low-lying zones are flat and have distinct shorelines. It looks as though methane rivers feed vast lakes. According to Martin Tomasko, leader of the *Huygens* camera team, "We saw Earthlike processes with very exotic materials. There is evidence for flowing fluids."

The marked color differences between the high ground, the river channels, and the lakebeds caused a stir. Although the solar energy hitting Titan is barely 1 percent as intense as on Earth, it is strong enough to trigger

Artwork from NASA shows what Jupiter might look like when seen from the icy surface of Europa. ⋀

A THEORETICAL MISSION

Europa is such an enticing target, it is the subject for an unusual "thought experiment" created by automotive designer Evan Twyford, a young veteran of several NASA pressurized rover prototyping experiments. Twyford's crowd-sourced space project, Objective Europa, is never intended to leave the ground. Instead, it serves to focus the minds of researchers, from student level to aerospace career professionals, as they investigate the theoretical possibilities of sending a crewed submarine-spaceship through the Europan ice crust and into its hidden liquid ocean. This is not just science fiction playfulness. Speculating about a mission of this kind lays the intellectual groundwork for real future space projects.

In July 2005 NASA's *Deep Impact* craft watched as a smaller probe deliberately smashed into comet Tempel 1. ◄

Technologists in new space companies are investigating how to extract mineral wealth from asteroids. ◄◄

chemical reactions in the upper atmosphere, breaking down methane to create a haze of complex hydrocarbons. These fall to the ground when it rains, leaving dark deposits. Titan is rich in the basic organic building blocks of life. The downside is that this is an unimaginably cold place. There is no possibility of liquid water, and the only oxygen available is tied up in hard ice. It is extremely unlikely that any life exists here. Even so, the chemical precursors on Titan promise to be almost as exciting as life itself.

The search for life beyond Earth has, of course, its shadowy side: the search for death. Of all the planets in our solar system apart from ours, one in particular carries a stark warning about what happens when environmental dynamics run out of control. Inappropriately named for the ancient goddess of love, Venus is an unfriendly world permanently clouded by a dense carbon dioxide atmosphere. None of its surface features are visible through conventional telescopes. Radar scans reveal energetic volcanism and mountains deformed by heat. The atmosphere traps energy from the Sun to such an extent that a runaway "greenhouse effect" has taken hold, making

this a hellish place where ground temperatures are high enough to melt lead. In December 1970, the Soviet Union achieved a successful landing of a robotic spacecraft on Venus. The lander survived for about twenty minutes before succumbing to the intense heat and pressure. It lasted just long enough to capture the first pictures ever transmitted from the surface of another planet.

Icy Wanderers

Our solar system is surrounded by a swarm of small, icy bodies known as comets. They contain clues about the earliest formation of our Sun and its planets. Most comets roam the Oort cloud (named after the Dutch astronomer Jan Hendrik Oort, who predicted the cloud's existence half a century ago). This is a vast spherical zone far beyond the orbit of Pluto and extending one third of the distance to the next nearest star, Proxima Centauri. Some comets interact with our solar system at regular intervals, drifting in a disk-shaped realm that lies beyond the orbit of Neptune but is closer to us than the Oort cloud. This realm is known as the Kuiper belt. When dust grains from a comet's tail

collide with the Earth's atmosphere, they burn for a few seconds and then are gone, causing meteor showers as insubstantial as sprites. A well-known display, the Perseid shower, peaks in intensity around mid-August. Comet Swift-Tuttle swept across the Earth's orbit in 1862 and left slight traces in its wake. Eventually they will dissipate.

Comets are probably the smallest accretion products that can occur as a solar system begins to take shape. If we could capture one, we might see the tiniest seeds from which a star or a planet might have formed. Some comets have been observed at close range by robot space probes. The main constituents are water ice, frozen carbon dioxide, and silicate dust grains, but many also contain simple organic molecules, such as ammonia, formaldehyde, and other hydrocarbons. Comets falling onto the young Earth probably contributed some of the building blocks for life. These tantalizing targets orbit the Sun at vast distances and are much more difficult to study than near-Earth asteroids, yet they are among the most scientifically rewarding objects that we could conceive of visiting.

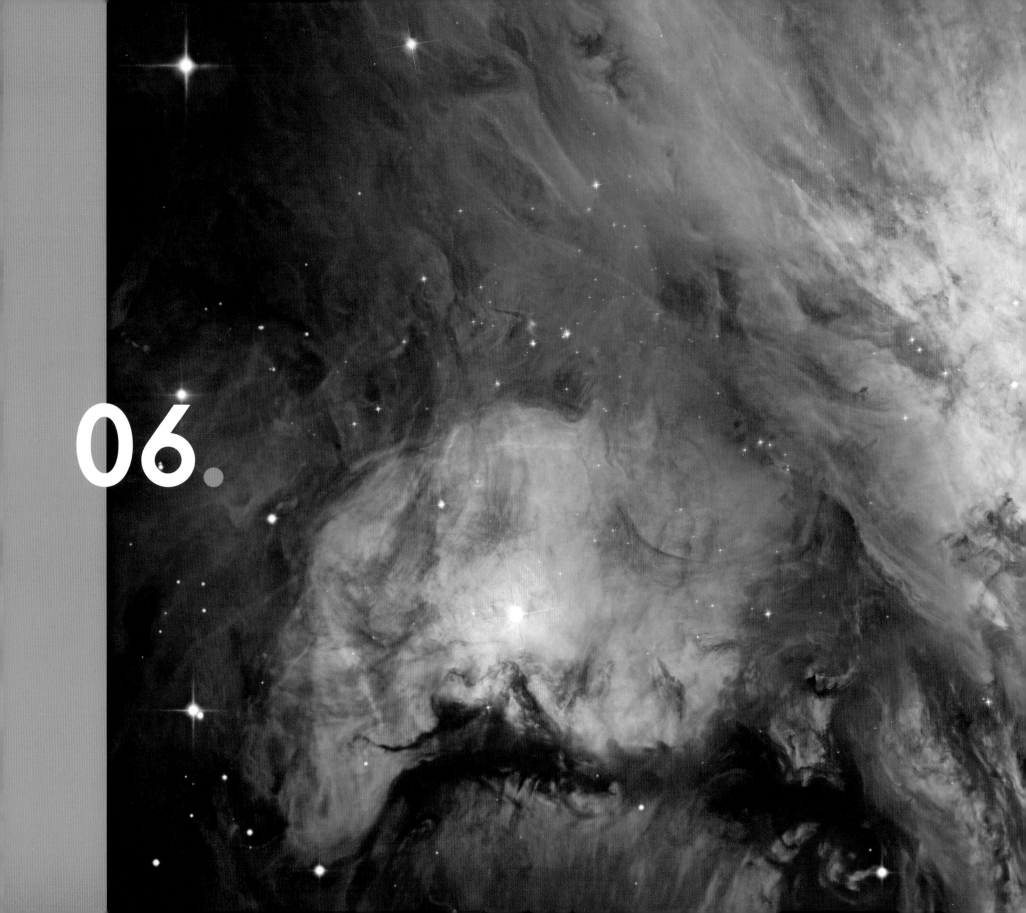

06.

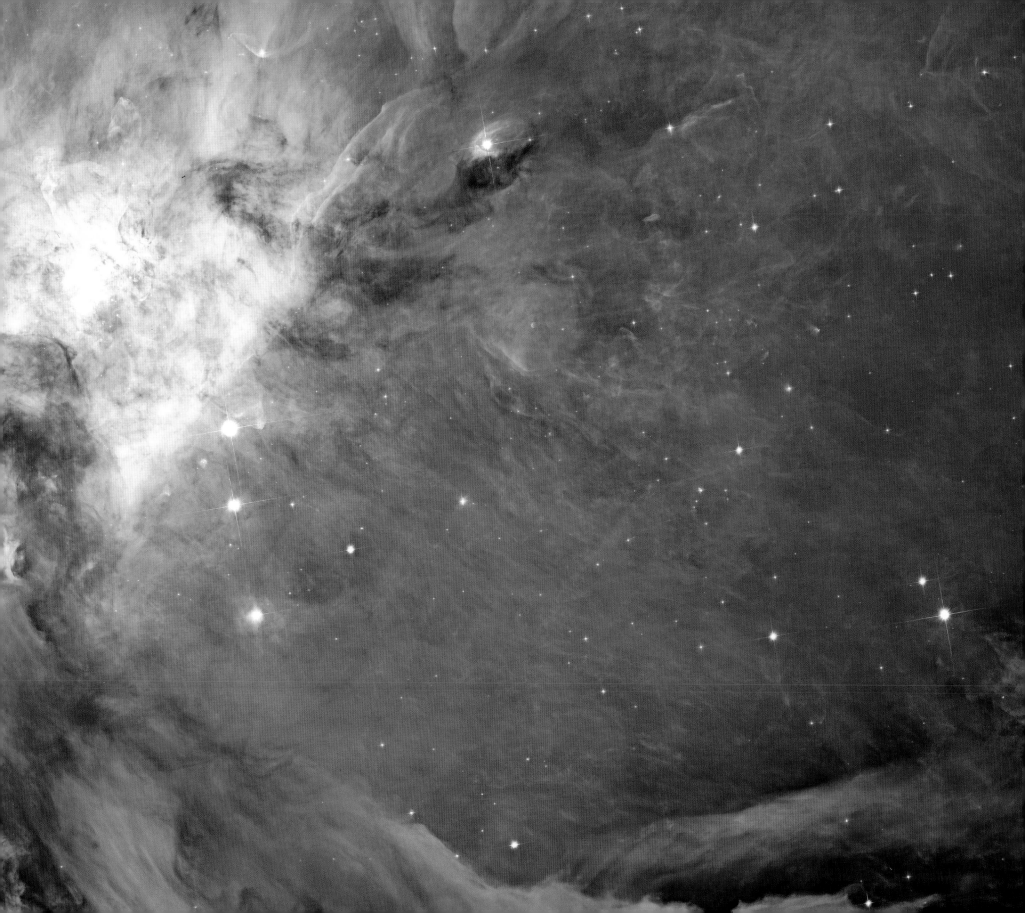

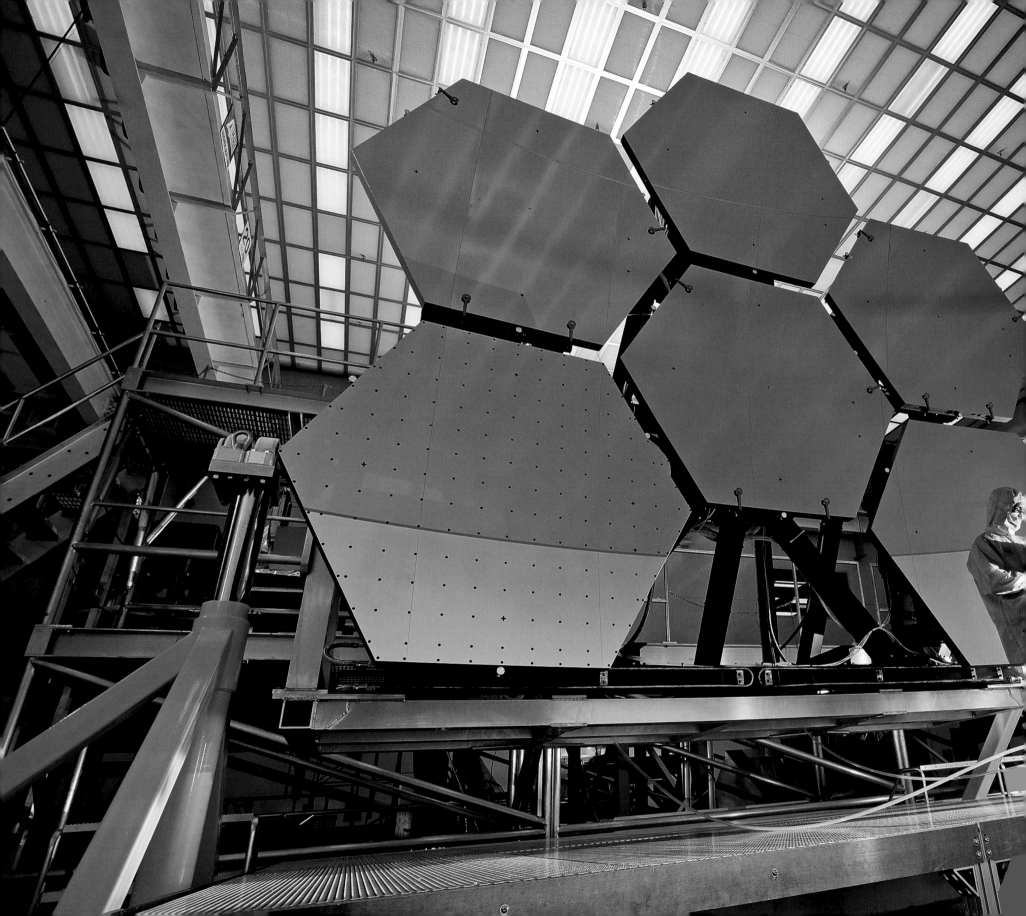

Traveling to distant stars is not just a science fiction dream, but while we wait for new starships to be invented, Earth-based telescopes will bring the stars closer to us.

ACROSS THE
GULF OF STARS

The late Douglas Adams, author of *The Hitchhiker's Guide to the Galaxy*, famously wrote, "Space is big. You just won't believe how vastly, hugely, mind-bogglingly big it is. I mean, you may think it's a long way down the road to the chemist's, but that's just peanuts to space." One day we might be able to explore neighboring star systems, but it will be a while before that happens. All we can do at the moment is crawl slowly across our little solar system. Orbiting astronomical instruments will always be crucial to our understanding of the deeper cosmos. For example, the Hubble Space Telescope, launched in April 1990, completely revolutionized our perspective. But what will come next for astronomy?

Expected to launch in 2018 on a European Ariane 5 heavy-lift rocket, the James Webb Space Telescope (JWST) is named in honor of NASA's chief administrator during the 1960s development of Project Apollo. The core instrument is an infrared telescope with a twenty-one-foot-diameter primary mirror mounted on top of a sunshade the size of a tennis court. The supporting spacecraft, with gyroscopic

stabilization systems, communication equipment, and solar power panels, juts out from the bottom of the sunshade.

Pam Sullivan, lead manager during the construction of JWST's Integrated Science Instrument Module (ISIM), explains the requirement for such a large primary mirror. "It will be looking for very distant and faint galaxies, and often will pick up no more than a single photon of light per second from a target, so we want to catch as many photons as we can." This extreme need for sensitivity requires JWST to operate in deep space, far from the interference of Earth's noise and heat. It will orbit the Sun along the plane of the ecliptic (the disk-shaped realm occupied by most of our solar system's planets) while simultaneously making a smaller orbit oriented perpendicular to the ecliptic and centered on a special zone known as Lagrange 2 (L2), about 930,000 miles from Earth. Here, the combined gravitational influences of the Sun, Earth, and Moon will allow the distant telescope to keep pace with the Earth's orbit around the Sun. To visualize this, imagine a dot of ink on the end of your forefinger. That's the JWST. Now stretch out your arm. Your head represents the Sun. The button on your shirtsleeve is the Earth, and the base of your forefinger is the L2 point. Turn your body while moving the tip of your finger in a little circle, and you have some idea of the complicated path that JWST will fly. The gravitational balances will enable the spacecraft to hold its position for many years, using the bare minimum of thruster burns.

Hubble-style astronaut repair missions are unlikely because of the telescope's great distance from Earth. For the first years of operation at least, JWST will have to fix its own problems, if any occur. The huge primary mirror is divided into eighteen separate hexagonal panels, each about the size of the door on a truck's cab. Each panel can be moved by remote control to adjust the focus. Everything else has to work without fail, too, first time out of the box. Like a metamorphosing insect in its chrysalis, JWST sits inside the nose cone of the Ariane rocket during launch, and then, once in space, it deploys in a complicated mechanical sequence of unfurling sunshade blankets, extending boom arms and fold-down mirrors. Sullivan is unconcerned. "It looks tricky, but Northrop Grumman, our main contractor for the spacecraft, are very experienced with these kinds of large deployable structures."

It is essential to shield JWST from the heat of the Sun, because infrared light and heat energy are closely related. The huge five-layered thermal blanket blocks out the heat and the glare, and the ambient temperature on the shadowed side is maintained at just a few tens of degrees above absolute zero. JWST's instruments have to be kept this cold so that they are not swamped by false signals. They have to be screened even against the tiniest traces of warmth from the spacecraft's own electronics. The five membranes of the sunshade are made from Kapton, a patented heat-reflective material as thin as a human hair. The two lowermost surfaces are coated with silicon, giving the sunshade's underside a distinct pink hue.

JWST's relatively wide field of view will encompass thousands of extremely distant astronomical objects, each giving off a slightly different light spectrum. The challenge

A set of mirror panels for the JWST is prepared for tests. ≪

A full-scale model of the JWST on display in Austin, Texas in 2014. ❯

"HUBBLE LET US SEE TEN BILLION YEARS BACK IN TIME. THAT'S THE ADOLESCENCE OF THE UNIVERSE. WHAT WE NEED TO SEE IS ITS CHILDHOOD"

JWST Lead Scientist Dr. John Mather, 2010

The multi-layered shields protect the JWST's main mirror from unwanted solar glare. ❯

THE SCIENCE OF JWST

The Integrated Science Instrument Module, or ISIM, is the heart of the JWST: the payload that makes the entire mission worthwhile. Its four main instruments are:

Mid-Infrared Instrument (MIRI) built by the European Space Agency (ESA), and NASA's Jet Propulsion Laboratory (JPL): multi-purpose camera for studying distant stellar populations, hydrogen clouds, and faint comets or Kuiper Belt objects at the fringes of our Solar System.

Near-Infrared Camera (NIRCam) built by the University of Arizona, Tucson: detects light from the earliest stars and galaxies in the process of formation.

Near-Infrared Spectrograph (NIRSpec) designed by ESA and NASA's Goddard Space Center: contains the special shutter arrays capable of observing more than 100 targets simultaneously.

Fine Guidance Sensor (FGS) from the Canadian Space Agency: contains a filter that can select extremely specific wavelengths of light and doubles as a star tracker for orienting JWST.

is to isolate and distinguish their spectra without having to repoint the entire telescope from one target to another. The solution is an array of hinged shutters, deep inside the ISIM, that can be individually opened or closed. First, a magnetized bar sweeps across the array and pulls all the shutters open. Then electrical commands are sent down specified rows and columns, and where these intersect, the chosen shutters are held open to let in light from specified targets, while all other shutters are closed. Thousands of star spectra can be analyzed simultaneously by this system.

There are sixty-two thousand shutters arranged in what looks like a waffle grid—if you use a microscope. Each shutter measures just two hundred microns across: about

A European Ariane V rocket will launch JWST into space, stowed (below) in a folded configuration. ⌄ ❯

the width of four human hairs. The entire array is less than two inches wide. JWST's instruments require four shutter arrays, each capable of surviving thousands of open-shut cycles. Mary Li of NASA's Goddard Space Flight Center, led the shutter fabrication team. "It's like coordinating the individual dusty scales on a butterfly's wings, except that none of the scales can be allowed to fall off. If one shutter breaks or jams open, all the ones alongside it become useless, because light rays from adjacent targets can't be allowed to interfere inside the detectors. We can only lose a few dozen individual shutters throughout the mission."

These shutters are just one of several new technologies that NASA has developed for JWST. For

instance, the mechanical assemblies to adjust the mirrors have never been flown before. There is also a new refrigeration system that pumps helium around the instruments, then radiates away any last traces of heat. This is all a huge challenge, but Sullivan is confident that "everything is on track, and when JWST gets into space, it will take us closer than we have ever been to the origins of the universe. And that's the day we are all working towards."

Astrophysicist John Mather shared the 2006 Nobel Prize in Physics with Professor George Smoot for their work with the Cosmic Background Explorer satellite, or COBE, the 1989 mission that charted the microwave background radiation associated with the Big Bang. Mather is JWST's

> ## "THE UNIVERSE IS A BIG PLACE. IF IT'S JUST US, THAT SEEMS LIKE AN AWFUL WASTE OF SPACE"
>
> Carl Sagan, 1985

Senior Proje ct Scientist and is as passionate about this new project as he was about COBE. "We think we're pretty smart, and that we have an accurate idea about the universe," he says. "We also think that so-called 'dark matter' provides most of the mass, and 'dark energy' causes the expansion of the universe to speed up. Then there's ordinary matter, the small proportion of stuff that we can see right now in our instruments. So we have these all these brilliant theories. Now we want to see if any of them are true. Also, of course, we want to see the difference between the galaxies that we can see today, and the galaxies that existed when the universe was very young. Hubble lets us see thirteen billion years back in time, at best—and it's a pretty good 'best', but we're dealing, still, with the adolescence of the universe. What we need to see is its childhood."

Infrared Vision

Hubble's breathtaking images are derived mainly from the visible and ultraviolet regions of the spectrum. JWST will operate primarily with infrared frequencies. There are plenty of good reasons for this choice. Space is suffused with dust clouds, the ancient wreckage of stellar explosions and other cosmic events. These are also the essential building materials for new stars, but the downside from our point of view is that the clouds block out much of the light from objects behind them. Fortunately the longer wavelengths at the red end of the spectrum can pass through.

The other factor is redshift. In 1929 the astronomer Edwin Hubble demonstrated that all galaxies are flying apart from each other, causing the light from distant galaxies to stretch into longer wavelengths at the red end of the spectrum by the time that we see it in our telescopes. The further away a galaxy is, the younger it is, because of the time taken for its light to reach us. Hubble can see galaxies dating from as long as thirteen billion years ago, but JWST will collect extremely red-shifted light dating from just a few million years after the Big Bang. Closer to home, JWST will also make images of planets orbiting nearby stars within our galaxy and try to determine whether any of them might be capable of sustaining life. In our more immediate neighborhood, the telescope will investigate comets and Kuiper belt objects at the fringes of our solar system. Looking at these distant objects through nearby telescopes is one thing. Sending humans to visit them represents quite another magnitude of problem altogether. It's time to talk about some mind-boggling numbers.

The Stars Like Dust

Our solar system drifts at one end of a spiral arm in a fairly typical galaxy within a universe so vast, there is no easy way for us to grasp the distances involved. The Earth orbits the Sun at a distance of ninety-three million miles. This is a convenient yardstick for comparing distances in and around our solar system. The Sun-Earth distance is known as an astronomical unit (AU). To get an impression of what an AU really means, think of the Sun as a grapefruit and the Earth as a blueberry about fifty feet away from it. Pluto's average distance from the sun is about forty AUs. (Think of Pluto as a grain of rice one-third of a mile away from the grapefruit Sun.)

The outermost realm of our solar system, where lonely comets lurk, is somewhere around one hundred thousand AUs in diameter. Beyond that cold and distant realm, even the AU loses its descriptive power as a measure of distance. We have to turn, instead, to the fastest entity in the universe: light. Light travels one AU in eight minutes. When measuring distances in the wider cosmos, we need to think in terms of how far light travels in years, decades, and millennia. The speed of light in a vacuum is 186,282

miles per second. In one year, an unimpeded photon travels 5,865,696,000,000 miles. Distances on this scale become unwieldy when written down as simple numbers. For convenience, astronomers use the term *light year*. The nearest star to our Sun is Proxima Centauri, 4.2 light years away. Imagine brilliant, glowing oranges separated from each other by the distance between New York and Chicago, or between Paris in France and Glasgow in Scotland. This gives a rough idea of the mind-numbing voids between stars that are essentially right next door to each other in galactic terms.

Our local galaxy, the Milky Way, contains at least two hundred billion stars. From one side of the galaxy to the other, the distance is around one hundred thousand light years, and its densely populated central core of stars is about thirty thousand light years deep. The light that we capture in our telescopes from stars at the galaxy's perimeter was emitted while Neanderthals still roamed the Earth. Astronomy is a window into the deep past.

Shaped by the collective gravitational forces of all the stars, gas, and dust clouds within them, galaxies usually adopt the form of spirals or ellipses. A spiral galaxy bulges in the center, where most of its stars are concentrated. The rotation of the galaxy causes outer stars to be dragged along in the gravitational tide as spiral arms. The Milky Way is a spiral galaxy, and it turns a full circle once every 220 million years. Our sun, residing in one of the outer arms, has traveled around the galactic center twenty times since our

solar system's birth 5 billion years ago.

A number of dwarf galaxies and diffuse clouds of stars are gravitationally linked with the Milky Way in a vague collection known as the Local Group. The nearest major spiral galaxy similar to our own is Andromeda, 2.5 *million* light years away from us. Sunlight reaches Earth in about eight minutes, already spanning in that brief time a distance more than the human mind can easily grasp. No human-scaled idea of any kind can encompass a distance so vast as 2.5 million light years—and that's also just nearby in cosmic terms.

Galaxies are distributed unevenly in space. Some drift in splendid isolation with no close companions. Others form strange pairs, orbiting each other, or in some cases, colliding and mingling. It seems incredible, but collisions between stars are extremely rare because of the inconceivably vast distances that separate them. Many galaxies are found in groups called clusters, typically millions of light years across. A cluster may contain from a few dozen to several thousand galaxies. Clusters, in turn, are grouped in superclusters. On even grander scales, galaxies tend to accumulate in a tangle of filaments surrounding relatively empty regions of the universe, known as voids. One of the largest structures ever mapped is a network of galaxies known as the Great Wall. This structure is five hundred million light years long and two hundred million light years wide.

The universe—that's to say, everything in space

The Atacama Large Millimeter/sub-millimeter Array (ALMA) radio telescope in northern Chile. ‹

The upper stage of a robotic Daedalus starship fires its fusion-powered engine. ›

that we can observe with our instruments—contains somewhere between one hundred billion and five hundred billion galaxies. There are probably more galaxies in the universe than there are stars in the Milky Way. Light from the most distant galaxies that we can observe has taken nearly thirteen billion years to reach us. That light first shone when the universe itself was still young. Because of these mind-boggling expanses of space and time, it is extremely unlikely that we will travel to another galaxy. However, we might be able to make some baby steps across our own.

Starship Troopers

In 1978 the British Interplanetary Society in London published a detailed specification for Daedalus, an interstellar robot probe designed by aerospace engineer Alan Bond. Its mission would be to fly to Barnard's Star, the third-nearest star to Earth, just under six light years away. Since Daedalus would be capable of achieving velocities of over eighteen thousand miles per second (around 10 percent of light speed), the journey would take less than fifty years.

Daedalus was conceived as a two-stage nuclear pulse rocket that creates thrust by generating small nuclear explosions in a rapid series of pulses. Tiny pellets of deuterium and helium-3 are injected into the propulsion unit and bombarded with electron beams, triggering nuclear fusion. The small yet powerful bursts of energy are repeated at a rate of two hundred and fifty per second, using strong magnetic fields to steer the resulting exhaust plume of superhot plasma gas.

Daedalus fires these pulses for two full years, consuming forty-six thousand tons of fuel, until the time comes to jettison the spent first stage, so that the ship can continue under power of its smaller, lighter second stage. During this stage, it fires off an additional four thousand tons of pellets until the craft reaches an extraordinary velocity of around one-tenth of light speed, after a further two years' gentle acceleration. Despite these unimaginable speeds, Daedalus is not streamlined. A huge disc-shaped buffer plate at the front end of the ship absorbs the high-energy impacts from cosmic dust particles.

All these processes are monitored by an advanced artificial intelligence, accompanied by a small fleet of onboard drones, each capable of flying independent local

The flat disc-shaped shield at the front of a Daedalus-class starship protects it against minor impacts. ◄

missions within Barnard's Star's local space. Daedalus never returns to its home world. Instead, it communicates its discoveries by radio. The data would take six years to reach Earth. The project scientists could expect their first results just fifty-six years after the starship's launch.

NASA believes that in another century or so, we will have spaceships capable of one-quarter of light speed, but these are likely to be very small and staffed by electronic, rather than human, minds. Tiny probes, rather than vast multiton starships, may be best for the job. Even with known physical limitations in mind, a slow, heavy, crewed version of Daedalus could fly to the nearest stars in a matter of centuries. Routine traffic among close-knit star systems could become common. For human astronauts, hundreds of years might seem too long a time to spend shut away inside a spaceship, starring at the walls and eating frozen meals. Individuals would have to commit their working lives to a mission—and the lives of their children, specially reared aboard the craft in order to complete the task. This sacrifice might be made worthwhile by building slow yet very large, luxurious ships capable of accommodating substantial communities of people: lovers, families, parents, and children. The descendants of those who set off on the voyage would be the ones who finally arrive at the target star system.

Perhaps these problems will become less extreme when the duration of an interstellar mission is considered as a proportion of a very long human lifespan. Mariners of previous generations were willing to tolerate sea voyages lasting many months, and sometimes several years. In the late eighteenth century, Captain James Cook and his crew made a series of exploratory voyages across the Pacific Ocean, with round trips typically lasting several years. At that time, in the 1760s and 1770s, the typical lifespan for an ordinary sailor was usually shorter than forty-five years. At least half of that life would have been spent at sea. Now consider the possibilities of some future century. Medical technology will almost certainly produce significant increases in average human lifespans, at least for a privileged minority. Already, we can extend the lives of mice and even reverse their aging processes.

If humans can stay alive and healthy for two hundred years—barely doubling our current maximum—then, for motivated space explorers, a long starship voyage might seem acceptable, particularly if the tedium and isolation could be eased through companionship and suitable pharmacology (sex and drugs) or suspended animation (hibernating in deep freeze). On coming home after a century or more in space, there is even a chance that explorers from some future human generation would see their relatives again. Even so, the challenges of interstellar star flight are daunting. We may choose to let tireless, artificially intelligent avatars take on much of the discomfort of initial explorations on our behalf.

Robot Ambassadors

So far, we have looked at future scenarios for humans in space and the prospects for new missions within our solar system. In half a century and more of rocket exploration, astronaut explorers have reached no further than the Moon. We dream of colonizing Mars or mining asteroids. We may very well accomplish these tasks in the coming generation. Meanwhile, we depend on machine ambassadors acting on our behalf. As in some science fiction scenario, clever robots already are scattered widely across our solar system. All deep space probes and planetary landers are capable of operating semiautonomously, enabling them to cope with the long time lags between radio commands from distant Earth. Some machines, such as NASA's Mars rovers, also boast certain kinds of mechanical flexibility, such as pan-and-tilt camera arrays, steerable wheels, soil sampling arms, chemistry labs, and other useful capabilities.

The public sometimes struggles with the idea that these clunky boxes are definable as robots, but no one can be in much doubt about NASA's latest nonhuman space explorer. At first glance it looks just like an astronaut in a space suit. Its head is shaped somewhat like a motorcycle helmet, and its torso and arms, tipped by elegant hands with slender fingers and opposable thumbs, look very like our own. Even at second glance, it seems at least as lively as a *Star Wars* character in an armored mask. Boba Fett springs to mind, although a ruthless bounty hunter may not be everyone's idea of a comforting comparison.

Robonaut, the first humanoid robot ever sent into space, was delivered to ISS aboard shuttle Discovery as part of the payload for the STS-133 mission in February 2011. It was stowed for launch inside the European-built Leonardo cargo supply module (which is now docked permanently).

This Robonaut is, in fact, a second-generation machine, based on a prototype developed by NASA and the Defense Advanced Research Projects Agency (DARPA) over the last decade. It is now coming to fruition via a joint project between NASA, General Motors (GM), and Houston-based Oceaneering Space Systems. Special legislation allows all three partners to protect their joint intellectual copyright in a technology that should have far-reaching implications on the ground as well as in orbit. Alan Taub, GM's recently retired Vice President for Global Research and Development, oversaw his company's contribution to Robonaut. "For us, this is about safer cars and safer manufacturing techniques right here on the ground."

It is also about the future of the American auto industry as it looks toward new and post-petroleum technologies. Assembly lines in Japan and Europe harness the productivity of the world's most advanced robotic assembly systems. GM wants to see "advanced robots working together in harmony with people, building better, higher quality vehicles in a more competitive environment." Robonaut is a high-profile testbed for some of these ambitions. NASA's priorities, of course, slightly different. Robonaut will set the tone for new working relationships between humans and machines in space. One day perhaps, humanoid machines such as this will explore worlds too hostile for humans to visit. They will report back tactile, auditory and visual experiences such as a human might experience, if only it were possible for a fragile member of our species to share the voyage. Jovian and Saturnian moons, for instance, may be too swamped by radiation to allow astronaut missions. Robonaut and its descendants could represent us by proxy.

Robonaut is designed to look and behave much like a space-suited human, with no unexpected protuberances or sharp edges. When it drifts in and out of the limited peripheral vision afforded by an astronaut's helmet visor, it seems perfectly familiar. Just as importantly, its physical gestures mimic the speed and range of behaviors that an astronaut would expect from a human colleague. According to Robert Ambrose, chief of the Software, Robotics and Simulation Division at JSC in Houston, "it operates at a speed and scale similar to us." Strictly speaking, no robot yet built can truly meet that description. What NASA means by "similar to us" is defined

" SPACE ONLY SEEMS LARGE. FOR US, IT IS CONFINING. DESPITE THE SIZE OF THE UNIVERSE, THE IDEA OF SPACE TRAVEL GIVES ME CLAUSTROPHOBIA "

U.S. journalist and critic Stanley Kaufman, 1968

SLEEPING THROUGH SPACE

Astronauts on long space voyages could be placed into suspended animation, a kind of cold storage during which their breathing, heartbeats and general metabolisms are reduced to the lowest possible limits consistent with the preservation of life. From the crew's point of view, a centuries-long mission might seem to pass in a moment of deep sleep. Unfortunately, the Earth would not be held in cold storage alongside the crew. They would have to cope with the fact that all their relatives back home would be aging and dying at the normal rate. A starship's outward journey might represent an acceptable emotional challenge to hardened space professionals, but any homecoming, even if technically possible, could prove unbearable.

A scene from *2001: A Space Odyssey* (1968)—three astronauts are unconscious in hibernation pods, while one is merely sleeping. ∨

ROBONAUT ON THE MOON

In 2009 a NASA study group wrote the brief for Project M, a proposed mission to land a Robonaut on the Moon in a thousand days flat, measured from the time of budgetary approval to the moment of touchdown. (M is the Roman numeral for 1,000). The delivery vehicle would be a small single-engined lander built by Armadillo Aerospace in Mesquite, Texas, one of more than a dozen companies now making rapid advances in the private rocket business. Armadillo has tested a prototype lander fueled with liquid methane and liquid oxygen. NASA's robonaut is nearly ready to fly. All that is needed is the funding.

A lunar version of Robonaut will operate in near real-time, once the operators on Earth have acclimatized their working habits around the one second or so of radio time lag in the command and feedback signals. Some of the designs show the robot walking on the Moon on two legs, but for the first experiments at least, bipedal locomotion may be too ambitious. Neither can we expect any poetic speeches when the robot takes its first look at the lunar terrain. It will be working alone and won't need to talk. Project M has not yet received the green light from NASA, but we can expect a wide range of robotic space explorers very soon, including several from the private sector.

Robonaut, the future space explorer: something not quite human but destined, one day, to become more than just a machine? ◀

in terms of the somewhat reduced dexterity, movement rate and performance of a human in a bulky space suit. Nevertheless, Robonaut is spookily impressive—and to be fair, its grasping skills and sensitivity to touch are markedly better than those of an astronaut, because its slender fingers have no need for thick gloves.

Fourteen motors in each forearm drive the hands, each of which has the proper complement of four fingers and an opposable thumb, ensuring compatibility with all the tools designed for astronauts' use. Robonaut can pick up a small metal washer with tweezers and has the strength to lift twenty pounds in weight. This may not sound like much, but it is more than sufficient for the inertial loads that it will encounter in microgravity. Robonaut's gold-plated head is articulated at the neck, allowing a similar freedom of movement to that of a human head. The torso is built from aluminum, with Kevlar and Teflon padding to protect it from fire and micrometeorite impacts. The torso and arms are covered with a fabric skin that secures the electrical wire harnesses and keeps dust away from the mechanical joints. An inner layer of foam padding absorbs the impact energy of minor and permissible collisions between Robonaut and its astronaut companions.

Finally we come to the lower extremities. The torso tapers neatly into the top of a single leg, with hip, knee, and ankle joints. The "foot" is an adaptor that clicks into various attachment points in the space station. Swapping between available handholds and leg attach points, Robonaut can clamber from one secure location to the next, covering ground, so to speak, as it makes its way from one end to the other of the sprawling complex, but without ever completely letting go.

Robonaut's hazard assessment routines are loosely

comparable to the famous "Three Laws of Robotics" proposed by science fiction writer Isaac Asimov more than half a century ago. In simple terms, an Asimovian robot has freedom of action unless the proposed action threatens to cause harm to a human. Similarly, Robonaut operates a hierarchy of subroutines, with harm avoidance at the top of the pyramid. Higher-level software monitors potential conflicts between safety parameters and assigned physical tasks before allowing or disallowing certain movements.

The Puppet Master

According to Robonaut's project manager, Ron Diftler, "Here's a robot that can see the objects it's going after, feel the surrounding environment and adjust to circumstances as needed. That's pretty human. It opens up endless possibilities." By this description, Robonaut sounds suspiciously smart. If its actions and intelligent behavior seem almost human, this is because they really are human. Despite its semiautonomous capabilities, it is essentially an avatar, driven by a human operator inside the space station or down on the ground. A stereovision helmet allied to a

force feedback endoskeleton and a pair of gloves studded with touch-sensitive (haptic) sensors enables the operator to see and feel Robonaut's environment. The system is an advanced form of what's known as telepresence.

Could it be that Robonaut and his kind will one day supplant humans in space? NASA insists that such an outcome would be the very opposite of the agency's intentions. Robots of this type are not intended as replacements for us, but to serve as companions that can carry out key supporting roles alongside us. Robonaut's potential next-generation upgrade, Valkyrie, is already undergoing tests. It is built around an existing set of hardware from Boston Dynamics (one of several robotics companies recently acquired by Google).

Ultimately the choice between human and machine exploration of space will become irrelevant. As Arthur C. Clarke famously observed, "One day we will not travel in spaceships. We will be spaceships." As long ago as 1958, the brilliant Hungarian American computer pioneer John von Neumann suggested that the march of technology "gives the appearance of approaching some essential

Previous spread: Robonaut (left) equipped for surface exploration, and its potential replacement, Valkyrie (right). ≪

A Robonaut prototype waves goodbye to a departing shuttle. ⌄

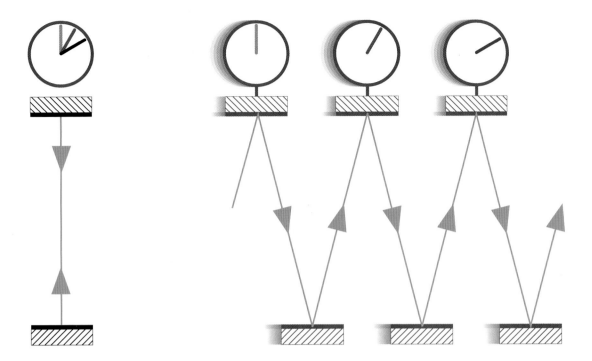

TIME DILATION IN ACTION

Imagine a clock that "ticks" by using pulses of laser light bounced between two mirrors inside a jet flying fast, but at constant velocity. The pulses bounce back and forth between the mirrors, one on the cabin's ceiling, the other on the floor. The pilot sees the laser beam firing straight up and down. However, if scientists on the ground could look inside the jet as it hurtles overhead, they would see light leaving one mirror and traveling diagonally to the next, because each mirror has shifted slightly forward by the time each pulse of light arrives.

From the ground's point of view, the pulses travel a long zig-zag path, rather than the short up-and-down one observed by the pilot. Therefore, when counted by a clock on the ground, the pulses take longer to complete each bounce than they do when measured by the pilot's clock. At the end of the experiment, the jet lands and its clocks are compared against the Earth's. Less time has passed on board the jet.

singularity in the history of the race, beyond which human affairs as we know them cannot continue." In 1965 Irving Goode, a similarly brilliant British mathematician, predicted "an intelligence explosion" trigged by "an ultra-intelligent machine that designs even better machines. The first intelligent machine is the last invention that man need ever make, provided it is docile enough to control."

At the University of California, Los Angeles (UCLA), an idea known as transhumanism took shape in the 1980s, just at the point where personal computers were becoming ubiquitous. It was possible to imagine a time, a few decades down the line, when biology and electronics might be merged, so that both brain and body could be augmented with additional powers of memory, intelligence, social connectivity, agility, and longevity. According to the transhumanist manifesto, we will become "better than well."

This sounds like rampant science fiction, until we remind ourselves of what's already out there. We have artificial limbs activated by nerve impulses. There's Darwinian software that designs its own improved successor code. 3D printers are capable of reproducing themselves. We are on the verge of wearable computing, from Google Glass to the iWatch. Already the distinction between the averagely wealthy human and his or her hardware is blurring. It's also obvious that certain kinds of virtual experience will become indistinguishable from real-world sensory impressions any day now, once we've broken beyond rectangular plasma screens and gone seriously three-dimensional. NASA's Robonaut and Valkyrie programs are baby steps along this road.

Space-Time

We think of space as having three dimensions: up-down, left-right, forward-backward. Objects, such as stars and planets, cats and dogs, occupy particular volumes of three-dimensional space. However, in Einstein's description of the universe, a fourth dimension has to be taken into consideration. When does something occupy space? Relativity wraps space and time into four-dimensional space-time. Time can be stretched out or compressed just like any other dimensional measure, such as length or width. The speed of light stays constant while space-time bends and stretches. The best way of illustrating this is to tell the story of two twins and a spaceship.

Imagine that two astronauts, thirty-five-year-old twins John and Jane, are the leading candidates for the first

This is what a starship might look like. The twin rings house what physicists today can only call "exotic matter." This creates the warped space field. ◄

mission to another star system, four light years away. Jane is eventually selected to make the round-trip voyage, and she sets off in a ship that travels very close to the speed of light. Allowing for a few years' acceleration during the outward journey and an equivalent deceleration at the end of the homeward leg, let's suppose that it takes ten years for Jane to make the round trip, as measured by the expensive digital watch that John wears at all times on his wrist.

John stays behind, helping to run mission control. When his twin sister finally returns and hugs him in celebration, John is startled by the digital readout on his sister's watch, even though his training should have prepared him for the shock. For him, ten years have passed, while only half that time seems to have passed for Jane. When her ship accelerated, or decelerated, she grew older at close to the normal terrestrial rate, but during the phases of her journey when she was traveling at close to the speed of light, time aboard her ship was compressed relative to time back in mission control. Jane noticed nothing unusual about her watch during her flight. For her, time seemed to pass as normal. It felt like five years, both according to her senses and her watch. Now she finds that she is forty years old while her twin brother, born on exactly the same day as her, is forty-five.

There should be a paradox here. Jane is entitled to think of herself as the stationary twin, motionless within her local environment (her starship) while the Sun and the Earth speed away from her. From her point of view, then, the people Earth surely should be the ones who stay young, while Jane gets older? In fact the broader cosmic frame of reference resolves the paradox. The Earth and the Sun continue on their cosmic course sedately while Jane's starship dramatically accelerates and decelerates within the wider cosmic environment, thereby incurring relativistic effects. We know this is true. No such fantastic flights to the stars are yet possible for human beings, but modern computer guidance systems routinely compensate for time dilation effects measured by digital clocks aboard fast-moving objects such as jet aircraft, rockets, and space satellites. The discrepancies between satellite and ground clocks amount to only a few fractions of a second, but they are real, and can cause navigational errors if left unadjusted.

Time is just one element of space-time. Relativistic effects also cause spatial dimensions to compress along

an object's direction of travel. According to Einstein's well-tested theories, If Jane's spaceship could travel at the speed of light, its length would shrink to zero. But no solid object can reach that speed. Relativity also shows that an object's mass increases with its velocity. As Jane's ship approaches close to the speed of light, it becomes so massive that no amount of energy available in the entire universe can accelerate it further. If it *could* actually reach the speed of light, its mass would become infinite! Light can travel at the speed of light because it has no mass. If it were possible to hitch a lift on a photon, no time at all would pass as it traversed from one side of the universe to the other—although, for external observers, its journey would be measured as taking over thirteen billion years.

Bringing the Stars to Us

NASA is working on an alternative approach. Rather than flying a ship *through* space, might it be possible to keep the ship stationary while warping the space around it? Recent astronomical discoveries strongly support the idea that the boundaries of space-time expanded many times faster than light during the first fractions of a second after Big Bang. It may be that, while light cannot move faster than a certain speed, space-time itself is not limited in such a way. In 1994, physicist Michael Alcubierre proposed a warp drive that could shrink the four-light-year gulf between Earth and the nearest star, Proxima Centauri, to just five months, without breaking Einstein's speed limit. Rather than flying *through* space, a stationary ship would create a "negative energy" field that would warp the environment around it, expanding space-time to the rear, and contracting space-time up ahead.

A future starship uses conventional nuclear thrusters to get clear of the Solar System before enaging its warp field. ⌃

The *Voyager I* probe from the 1970s: our first real interstellar craft. ❯

NASA researchers invite us to imagine a moving walkway such as can be found in many airports. Although a passenger hastening toward a terminal can only walk as fast as the maximum walking speed possible for a human, the walkway (space-time) can move considerably faster. The net effect is that the passenger reaches her destination in much less time than her walking speed alone could account for.

Exotic forms of energy would be needed to create such a warp: energy, what's more, in the dangerous form of anti-matter. Recent research indicates that perhaps a mere ton of anti-matter would suffice. Anti-matter annihilates ordinary matter on contact, creating a blast of gamma radiation. That precious ton of volatile fuel could be the single most dangerous object we ever create. However, if we can harness these energies safely and economically, they may open the road to the stars.

Warp drive is no longer pure science fiction. It is the subject of legitimate and detailed research, albeit at tiny laboratory scales for now. Extreme thinking today sows the seed for commonplace applications tomorrow. Remember, there was a time, three generations ago, when telephones were comparatively rare. Two generations ago, no one could see the need for more than one computer per city: no one, that is, apart from dreamers and visionaries who foresaw such things as Dick Tracy's wristwatch communicator. Most American homes have several powerful computers, some so small we do not even recognize them as computers. Even the wristwatch thing is commonplace now. If starships still seem like science fiction, bear in mind that we have already dispatched at least four of them.

The First Real Starships

In August 2012, after thirty-six years hurtling through space at approximately ten miles per second, the *Voyager I* space probe achieved an incredible milestone, which was announced to the world's press by lead scientist Edward Stone and his colleagues. At nearly eleven billion miles from Earth, the spacecraft is now "the most distant human-made object," and Stone is confident that it has escaped our solar system completely. "We hoped, all those years ago when we first launched, that this would happen, but none of us knew how long it might take to get there. We are in interstellar space for the first time."

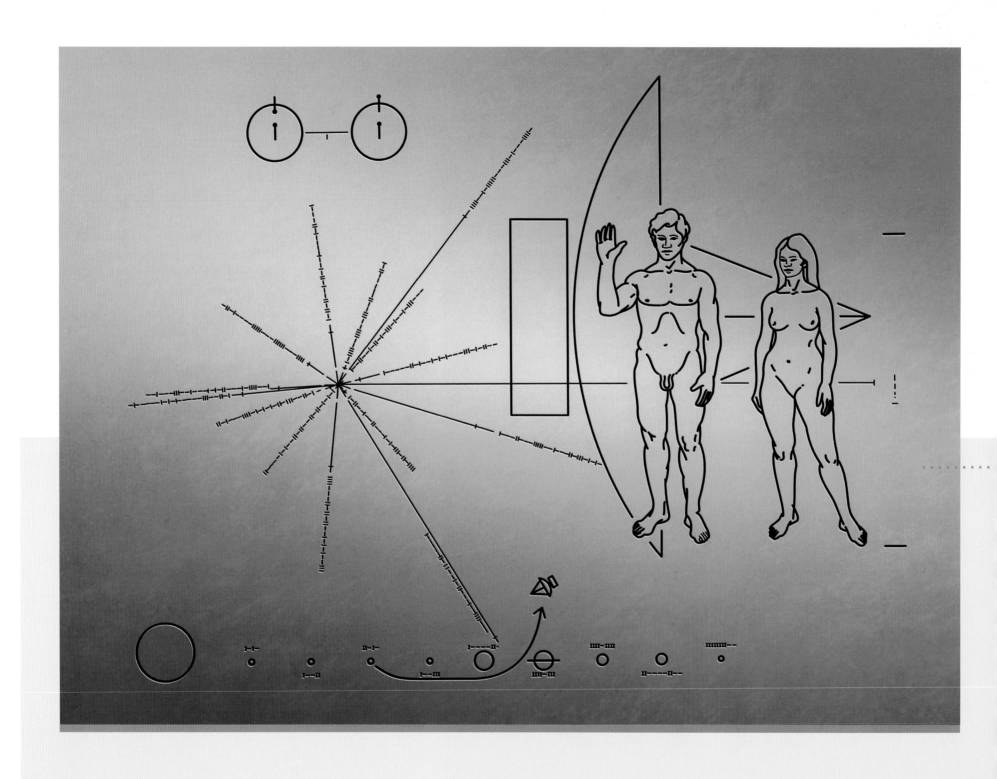

> # "VOYAGER IS THE MOST DISTANT HUMAN-MADE OBJECT. WE ARE IN INTERSTELLAR SPACE FOR THE FIRST TIME"
>
> Voyager lead scientist Dr. Ed Stone, 2013

"OUR" MESSAGE TO THEM"

This is the etched drawing affixed to the flank of each PIONEER space probe. Attitudes have changed since 1972, and we no longer share all the assumptions in the drawn symbols. Why is the man waving, and not the woman? NASA decided that if both humans held their hands up in greeting, the aliens would think that all of us walked around all day with one arm in the air. It made sense to show a variety of postures, although it never occurred to anyone in the 1970s that the womanw might do the waving. But there's one assumption that we still believe is correct. Math is a basic language that intelligent, technological aliens must share with us.

How can Stone be so sure this is the key moment? Space is not empty. The Sun's furious energies hurl the outermost layers of its atmosphere across our solar system. This constant energetic stream, known as the solar wind, is made up of plasma: atoms that have been shredded into electrically charged subatomic particles. When solar storms hurl particularly intense bursts of plasma in Earth's direction, satellites can be knocked out and terrestrial power supplies disrupted.

There comes a point, at the outermost edges of our solar system, when the pressure of this wind is depleted, and the opposing pressure of interstellar gas and charged particle streams from other powerful galactic sources of radiation answers back, forcing the solar wind to retreat. This is the heliopause, where the Sun's influence is finally defeated by the wider radiation influences and gravitational forces of the cosmos. According to Stone, "The solar wind creates a shock wave, pushing against the interstellar environment, and being pushed back in return. *Voyager* is now passing through that shockwave and into the realms beyond."

In about forty thousand years, *Voyager I* will approach the star Gliese 445 in the constellation Camelopardalis. Sister ship *Voyager 2* trails behind at 9.5 billion miles from the Sun. It may be another three years before it joins its twin on the other side of the heliopause, on its way to a distant rendezvous with Ross 248, a star in the constellation of Andromeda. Stone is amazed that the *Voyagers* are still functioning. "When we built them, the space age itself was young, so we had no way of knowing that any spacecraft could last that long. But they just keep going." The *Voyagers* will never relay to us any discoveries among the stars. They are running out of power. Their radioisotope generators convert heat released by the decay of plutonium-238 into electricity. These will cease functioning some time around the year 2025. Nevertheless, these are starships.

Powerless they may become, but the *Voyagers'* mechanical structures could survive for millions or even billions of years. It is unlikely that any alien intelligences will intercept them, but it's not impossible. If they do, special messages from us to them are waiting. Each *Voyager* carries a gold-plated metal disc, engraved with a spiral of dips and dents, using the best phonograph technology available in the 1970s, prior to the age of DVDs and laser discs. The outer protective case includes a diagram of how the stylus, included in the kit, can be used to "play" the record. Images of life on Earth, and a variety of natural sounds, such as surf, wind, and thunder, are encoded, along with the songs of birds and whales. Humanity is represented in fifty-five languages, and by greetings from President Jimmy Carter and the UN Secretary General of the time, Kurt Waldheim. Two more star-bound probes were launched half a decade before the *Voyagers*. *Pioneer 10* was the first spacecraft to fly by Jupiter in December 1973, coming to within eighty thousand miles of the gas giant's upper atmosphere. It is on a two-million-year-long journey to the red star Aldebaran in the constellation Taurus. A sister ship, *Pioneer 11*, is nine billion miles from the Sun.

Rectangular golden plaques on each *Pioneer* show where our species lives and its biological form. At top left of the diagram is a schematic of two hydrogen atoms bound up in a molecule of dihydrogen, or (molecular hydrogen). This is such a basic material in the universe, it should be known to any technological alien civilization. Its properties can be used as a yardstick for both time and physical length throughout the universe. That gets rid of the problem of metric, imperial, or other units of measurements that confuse us humans, never mind any aliens.

As a further size check, the binary equivalent of the decimal number eight is shown between tote marks indicating the height of the two human figures, standing next to the *Pioneer* spacecraft itself, which is also shown on the plaque. The radial pattern to the left represents the position of our Sun relative to fourteen pulsars (rapidly spinning stars that eject extremely regular energy pulses) and to the center of the galaxy. The binary digits on the other lines denote time. There is also a diagram showing *Pioneer 10*'s journey, accelerating past the largest planet in our solar system with its antenna pointing back to its origin on the third planet.

Would *you* understand the plaque, as drawn by people of your own species? Perhaps not at first, but intelligent aliens might dedicate their brightest minds to the task of unraveling its mysteries, so maybe they will have better luck. For now, the plaque is better understood (by us, at any rate) as a message to ourselves rather than to aliens. It signifies that we at least tried to head toward the stars, even while clashing ideologies and social problems afflicted

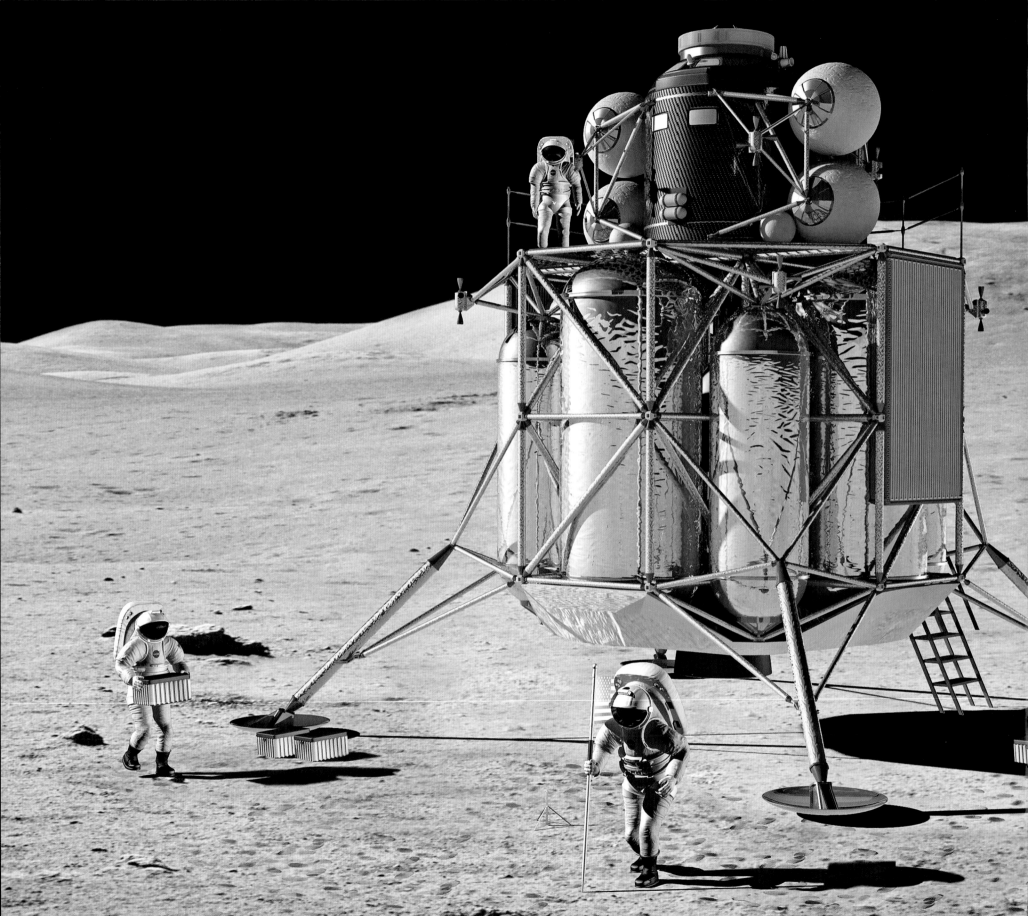

This concept for a lunar lander, conceived at the dawn of the twenty-first century, looks uncannily similar to designs from the 1950s. ‹

us on Earth. It is a technological cave painting, created as much because we felt like it as for any other reason: a portrait of ourselves as we would like to be seen.

There is one more human-made machine destined for the stars, but only after it has finished its work in this solar system. NASA's New Horizons probe was launched in January 2006 to study the dwarf planet Pluto. It will arrive in July 2015. Then it will investigate the Kuiper belt, a hitherto unexplored region inhabited by stray fragments of rock and ice left over from our solar system's formation more than four billion years ago. After that, the little probe will head toward the constellation Sagittarius.

A Spacefaring Civilization?

In a 1964 edition of the *Journal of Soviet Astronomy*, astrophysicist and extraterrestrial intelligence hunter Nikolai Kardashev devised the Kardashev Scale, a speculative ranking of technological advancement based on access to energy access. He proposed three levels: Type 1, which makes the fullest and most efficient use of energy on its home world; Type II, which harnesses the power of its local star to the fullest possible extent, and Type III, which exploits the effectively limitless resources of its surrounding galaxy. Freeman Dyson of Princeton's Institute for Advanced Study, a great and revered elder statesman of modern physics, thinks that we should be capable of reaching Type 1 status in a couple of centuries. Kardashev thinks that Type II ranking would take another thousand years. That seems like a long wait when measured against the scale of individual human lifetimes, but it is the merest blink of an eye in cosmic terms.

Even now, despite our crises of warfare and environmental hazards, optimists sense the beginnings of a Type I civilization on Earth. Arthur C. Clarke predicted the coming of a "Global Village." English is a commonly accepted language; the Internet is a catchall communication system, and we have a global economy and a global culture. Many people mourn the swamping of individual cultures amidst the irrepressible hum and buzz of Anglo-American pop culture, although the Global Village is evolving to take in Asian, Arabic, Indian, and countless other once-localized cultural influences, from food habits, to language, and of course, religion. Terrestrial civilization shifts and shimmers, cracks and mends, and the unifying glue is technology: the ever more efficient transportation of people, materials, foodstuffs, and above all, ideas.

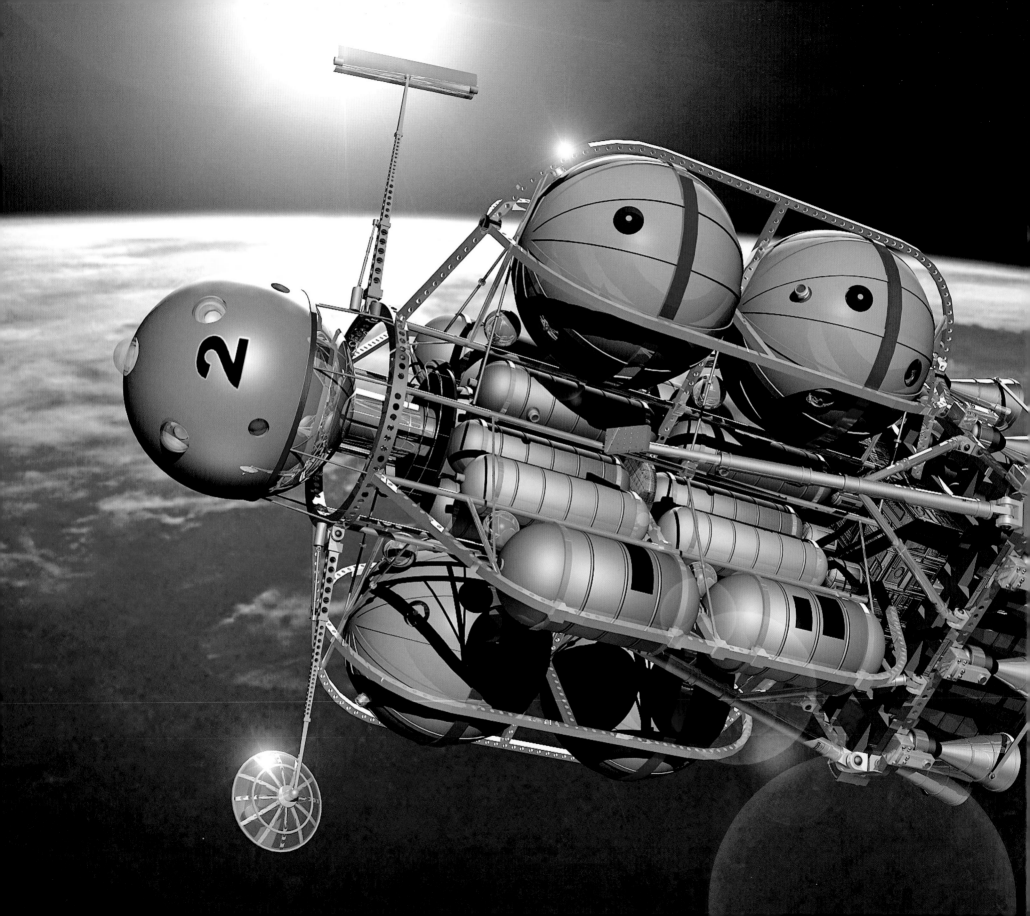

Computer graphics make Wernher von Braun's lunar landing ship design from the 1950s look modern. It can take many decades for such grand space projects to be realized. ‹ ^

" ONE DAY WE WILL NOT TRAVEL IN SPACESHIPS. WE WILL *BE* SPACESHIPS "

Arthur C. Clarke, 1973

Modern terrorism, border wars, international squabbles, and ethnic tensions—really they are all just reflections of the same fundamental problem. The greatest threat to the emergence of a Type I civilization is that we are forced to compete at local scales for natural resources: water, food, minerals, oil, gas, and living space. Limited resources create strife. For the time being at least, our current poor energy exploitation qualifies us only for Type 0 status. We derive almost all our energy in primitive ways: by inefficient combustion of fossil fuels and clumsy methods of secondhand heat extraction from lumps of uranium and plutonium, coupled with steam turbines, even as a vast source of energy beams down on us for free. The Sun emits energy in all directions. The Earth intercepts about one-billionth of the total output. Humans use around a one millionth fraction of that—which means we harness one million billionth of the Sun's total energy. Not very clever.

If we could free ourselves of energy constraints, we could eliminate all resource shortages, because energy makes anything possible. We could become a Type I society; albeit one still trying to survive in a dangerously small environment, on the surface of a small ball of rock, under constant threat of annihilation from viral pestilence, asteroid impact, or environmental carelessness. Even a Type I civilization is fragile. Physicist Stephen Hawking has warned repeatedly that our long-term survival depends on expanding into our solar system, and perhaps far beyond, just in case anything drastic happens to the one little patch of ground on which we stand at the moment. "I don't think the human race will survive the next thousand years, unless we spread into space. There are too many accidents that can befall life on a single planet. But I'm an optimist. We will reach out to the stars."

credits

I could not have obtained such a wide range images for this book without the help of many talented and generous individuals, organizations, agencies and companies. In particular I would like to thank Brandi Dean, Debbie Solomon, Burt Ulrich, Kimberly Henry, Mark Holderman and Allison M. Rakes at NASA for fielding my image requests with such patience. Similarly, Sean Blair and his colleagues in the media office at ESA delivered great materials.

Jessica Ballard at the Griffin Communications Group took care of many image requests related to a number of private companies and foundations shaping a new space age, including Virgin Galactic, Inspiration Mars and Sierra Nevada Corporation. Within those companies, I am grateful to Will Whitehorn, former President of Virgin Galactic, Kevin S. Kelley at the Boeing company, Heather Kelso at Lockheed Martin, Janea Laudick and Andrew F. Antonio at World View Enterprises, and Eva van Pelt at XCOR Aerospace. Many thanks also to Hannah Post at SpaceX.

Individual artists and illustrators created print-resolution digital files specially for this book. I am really grateful for the time and enthusiasm they all devoted to this task. Specifically, may I thank Adam Benton for his glorious depiction of a space habitat interior. John Frassanito & Associates are the "go-to" people for future NASA concepts. Eric Bernard & Wes Sargent at the Space Opera Society rendered a lunar base. Mark Rademaker created a fabulous starship, as did Frank da Silva at DMT Lab, while Prince Yemoh and Rachel Armstrong provided images of "living spaceships" from Project Persephone, and digital artist-engineer Jan Kaliciak "built" an asteroid mining craft. Adrian Mann brought Skylon to life and sent a giant robot starship to the Barnard star system.

The talented automotive designer Mark Twyford shared his visionary contributions to NASA's next-generation rover and multiple-role spacecraft projects. In conclusion, many thanks also to Erik Gilg, commissioning editor at Zenith Press, who essentially brought this book into existence.

picture credits

Contents & pages **vi–ix, 1–3** NASA/Sierra Nevada Corporation
4–5 NASA/Lockheed Martin/ESA
6–7 & **17–19** Virgin Galactic/Scaled Composites/Griffin Communications
8 NASA/Novosti
9–10 Bill Ingalls
11 Mark Shuttleworth *(top)*, ESA *(bottom)*
12–13 Anousheh Ansari/X Prize Foundation/NASA
14–15 XCOR Aerospace
16 Smithsonian Institution
18–19 Virgin Galactic
20–21 SpaceX *(main & top right)*, Boeing *(bottom right)*
22 NASA *(left)*, SpaceX *(right)*
24–25 StratoLaunch/Orbital Sciences/NASA
26 SpaceX *(left)*, Swiss Space Systems *(top left & right)*
28–29 Reaction Engines Limited/Adrian Mann
30–31 Sierra Nevada Corporation/NASA/Griffin Communications
32–33 Xinhua News Agency
34–35 & **42–45** World View Enterprises, Inc.
36 Lockheed Martin
37 Boeing
38–40 NASA
41 & **46–49** Red Bull Stratos
50–51 Adam Benton
52–53 Sony Pictures
54–59 NASA
60 SICSA
61 NASA/Mark Holderman
62 Bigelow Aerospace
63–65 NASA
66–67 & **70–71** NASA/National Space Society
68–69 Project Persephone
72–73 John Frassanito & Associates
74–75 Wes Sargent and Eric Bernard/Space Opera Society
76–79 & **82–87** NASA
80–81 Foster & Partners/ESA
88–89 Dava Newman/MIT
90–91 NASA/Mark Twyford
92–93 NASA/Bill Ingalls
94–95 NASA History Office
98–99 & **102–105** John Frassanito & Associates/NASA
100–101 & **107** & **111** NASA
108–109 The Mars Society
110 SpaceX
112–115 Inspiration Mars/NASA
116–120 & **122–123** & **127** NASA/Frassanito
121 Planetary Resources, inc.
124–125 NASA/JPL *(top left & top right)*, Mark Twyford *(bottom left)*
126 Jan Kaliciak/EOS Mars
128–137 NASA/HSTI/ESA
138–139 European Southern Observatory
140 National Radio Astronomy Observatory
141–143 Adrian Mann
144–145 Turner Entertainment/Stanley Kubrick Archive
146–150 NASA
152–153 Mark Rademaker
154 Frank da Silva/DMT Lab
155–156 NASA/JPL
158–159 John Frassanito & Associates
160–161 Terry Sunday *(main)*, NASA *(inset)*

index

Agena vehicles, 65

Akayima, Toyohiro, 26

Alcubierre, Michael, 154

Allen, Paul, 15

Altair, 3

Ambrose, Robert, 144

American Society of Civil Engineers, 7

Ansari, Anousheh, 11–12, 27

Ansari X Prize, 11–12

Antares rocket, 24

Ares rocket, 106

Ariane 5 heavy-lift rocket, 131–132

Armadillo Aerospace, 147

Armstrong, Neil, 31

Artsutanov, Yuri, 65

Asimov, Isaac, 150

Atlantis, 1

Australia, 37

Automated Transfer Vehicle (ATV), 5

Baturin, Yuri, 9

Baumgartner, Felix, 46

Bernal, J. D., 53

Bero, Chris, 92

Bigelow, Robert, 63

Bigelow Aerospace, 29, 61, 63, 65, 81

Bigelow Expandable Activity Module (BEAM), 65

Binnie, Brian, 12, 26

BioSuit, 88

Blaha, John, 32

Bond, Alan, 29, 38, 143

Boston Dynamics, 150

Branson, Richard, 15, 17

Breitling Orbiter 3 balloon, 41

Britain, 29, 38

British Interplanetary Society, 143

Bush, George W., 1–2, 75, 101

Bussey, Ben, 77

Canada, 92, 134

Canadian Space Agency, 92, 134

Carter, Jimmy, 157

Cassini space probe, 125

Challenger, 1

China, 31–33

Clarke, Arthur C., 53, 56, 65, 150, 159

Clementine space probe, 77

Club of Rome, 66

Columbia, 1–2

Commercial Orbital Transportation Services (COTS), 21, 26

Conrad, Pete, 65

Constellation project, 3

Cosmic Background Explorer satellite (COBE), 135, 137

Cygnus cargo supply vehicle, 24, 111

Daedalus, 143

Davis, Don, 68

Defense Advanced Research Projects Agency (DARPA), 144

Defense Science and Technology Organization (DSTO, Australia), 37

Diamandis, Peter, 11

Diftler, Ron, 150

Discovery, 1, 144

Dnepr booster rockets, 63

Dobrovolsky, Georgy, 8

Draco rocket thrusters, 23

Dragon capsule, 21, 23

DragonRider capsule, 23

DreamChaser, 30

Drexler, Eric, 68

Dyson, Freeman, 159

earth return vehicles (ERV), 103, 106, 109, 110

Endeavour, 1

Enterprise (NASA), 1

Enterprise (Virgin), 15, 17

Europe, 5, 8, 30, 75, 125, 131, 139

European Extremely Large Telescope (E-ELT), 139

European Southern Observatory (ESO), 139
European Space Agency (ESA), 30, 75, 125, 134
European Very Large Telescope (VLT), 139
Eve (Virgin), 15, 17

Falcon 1, 23
Falcon 9, 21, 23, 26
Falcon Heavy, 23, 26
Fei Junlong, 32
Fender, Donna, 62–63
Fine Guidance Sensor (FGS), 134
Foster, Norman, 18, 81
Fournier, Michel, 43

Gagarin, Yuri, 31
Garriott, Richard, 27
Gemini 11, 65
Gemini capsules, 65, 83–84
General Motors (GM), 144
Goddard Space Center, 134, 135
Goodall, Kirk, 99
Goode, Irving, 150–151
Gordon, Dick, 65

Habitation (HAB) living module, 106
Hawking, Stephen, 161
Hayabusa, 121
Henderson, Edward, 61
HL-20 mini shuttle project, 30
Holderman, Mark, 61
Hubble, Edwin, 137
Hubble Space Telescope, 1, 59, 123, 131, 137
Huygens, 125

In-Situ resource utilization (ISRU), 109
Inspiration Mars, 110–111
Institute for Advanced Study (Princeton), 159
Institute of Biomedical Problems (IBMP), 92
Integrated Science Instrument Module (ISIM), 131, 134, 135

International Astronomical Union (IAU), 120
International Space Station (ISS), 1–2, 5, 8, 9, 21, 24, 30, 52, 56, 59, 62–63, 65, 84, 87, 92, 144
Italy, 84

James Webb Space Telescope (JWST), 131–132, 134, 135, 137
Japan, 75, 121
Japanese Aerospace Exploration Agency (JAXA), 75
Jet Propulsion Laboratory (JPL), 33, 99, 134
Johns Hopkins University, 77
Johnson Space Center, 144
Jones, Brian, 41

Kardashev, Nikolai, 159
Kennedy, John F., 101
King, Alexander, 65
Kittinger, Joe, 41, 43, 46
Komarov, Vladimir, 8

L-1011 Stargazer (Lockheed), 24
Laliberté, Guy, 27
Lansdorp, Bas, 110
Lapierre, Judith, 92
Leland, Brad, 38
Leonardo cargo supply module, 144
Leonov, Alexei, 83
Li, Mary, 135
Liu Yan, 32
Lockheed Martin, 24, 38
Long March 2F rocket, 32, 33
Long March 5 rocket, 33
Long March 7 rocket, 33
Lunar Prospector, 77
Lynx, 15

Mars Colonial Transporter, 109
Mars Direct, 103, 106, 109
Mars One, 110
Mars Pathfinder mission, 99

Mars Society, 108
Martin Marietta, 101, 103
Massachusetts Institute of Technology (MIT), 66, 88
Mather, John, 135, 137
McKay, David S., 115
Meadows, Donella H., 66
Mid-Infrared Instrument (MIRI), 134
Mir space station, 32, 56
Multi-Mission Space Exploration Vehicle (MMSEV), 90
Musk, Elon, 21, 23, 26, 109–110

NASA (the National Aeronautics and Space Administration), 1–3, 9, 21, 26, 29, 38–40, 59, 61, 65, 68, 70, 75, 77, 79, 92–93, 99, 101, 103, 105, 111, 118, 119, 134, 144, 159
Nautilus-X (Non-Atmospheric Universal Transport Intended for Lengthy United States Exploration), 61
Near-Infrared Camera (NIRCam), 134
Near-Infrared Spectrograph (NIRSpec), 134
Netherlands, 110
New Horizons probe, 159
Newman, Dava, 88
Nie Haisheng, 32
Non-Atmospheric Universal Transport Intended for Lengthy United States Exploration (Nautilus-X), 61
Noordung, Hermann, 52–53
Nowak, Lisa, 93
Nuclear Engine for Rocket Vehicle Application (NERVA), 105

Obama, Barack, 3, 75
Objective Europa, 125
Oceaneering Space Systems, 144
Olsen, Greg, 9, 11, 27
O'Neill, Gerard K., 65–68, 70
Orbital Sciences Corporation, 24, 37, 111
Origins Spectral Interpretation Resource Identification Security Regolith Explorer (OSIRIS-REx), 121, 123
Orion, 2–3, 5, 30, 31, 61, 111, 119